ALL YOU SHOULD KNOW ABOUT CRIME PREVENTION AND PROTECTION—FROM THE MAN WHO *REALLY* KNOWS IT ALL

- What is the best alarm system for my home and car, and which locks give the most protection?

- How do I make doors and windows forced-entry-proof?

- What's my best move when an armed mugger demands my money?

- What con game is most likely to sucker me?

- How do I warn my child about crime without arousing unnecessary terror?

- What are the ten rules of crime prevention for the elderly?

- How do I thwart pickpockets, purse snatchers, and cheating cab drivers?

- How do I react when rape is threatened?

RAY JOHNSON is America's leading anti-crime consultant; his work for the Southland Corporation has reduced crime by some 60% in its chain of 7-Eleven stores. He has made numerous appearances on network TV, including *The Tonight Show* with Johnny Carson and *The Today Show*.

CARROLL STOIANOFF was formerly publisher and editor of *Pastimes*, the Eastern Airlines flight magazine. A freelance writer, he has written articles for leading magazines.

HOME FINANCE

☐ **THE REAL ESTATE BOOK by Robert L. Nessen. Revised edition.** A complete guide to real estate investment including information on the latest tax laws, mortgage tables, and a glossary of terms. "Indespensable"—*Forbes* Magazine   (140966—$4.50)*

☐ **THE COMPLETE ESTATE PLANNING GUIDE by Robert Brosterman.** A renowned lawyer and estate planning expert tells how you can set up a long-range program that will not only benefit your heirs, but safeguard your own financial security in your lifetime. Newly revised and updated to incorporate new tax law changes.   (623347—$4.95)

☐ **HOW TO BUY YOUR OWN HOUSE WHEN YOU DON'T HAVE ENOUGH MONEY! by Richard F. Gabriel.** A real estate specialist's guide to creative financing, dollar stretching, low-cash buying techniques, and dozens of home-buying, money-raising-and-saving strategies.
(129903—$3.95)*

☐ **GET YOUR MONEY'S WORTH: The Book for People Who Are Tired of Paying More for Less by Froma Joselow.** Learn all the ways—and how easy it is—to make your money buy more and grow faster today. If you're tired of your living standard going down when your income is going up, now is the time to start to *Get Your Money's Worth!*   (121325—$3.50)*

*Prices slightly higher in Canada

Buy them at your local bookstore or use this convenient coupon for ordering.

NEW AMERICAN LIBRARY,
P.O. Box 999, Bergenfield, New Jersey 07621

Please send me the books I have checked above. I am enclosing $_____ (please add $1.00 to this order to cover postage and handling). Send check or money order—no cash or C.O.D.'s. Prices and numbers are subject to change without notice.

Name_____

Address_____

City_____State_____Zip Code_____
Allow 4-6 weeks for delivery.
This offer is subject to withdrawal without notice.

Ray Johnson's Total Security

HOW YOU CAN PROTECT YOURSELF FROM CRIME

RAY JOHNSON
WITH CARROLL STOIANOFF

A SIGNET BOOK

NEW AMERICAN LIBRARY

PUBLISHER'S NOTE

This publication is designed to provide accurate and authoritative information
in regard to the subject matter covered. It is sold with the understanding that
the publisher is not engaged in rendering legal, accounting or other professional service. If legal advice or other expert assistance is required, the
service of a competent professional should be sought.

NAL BOOKS ARE AVAILABLE AT QUANTITY DISCOUNTS WHEN USED
TO PROMOTE PRODUCTS OR SERVICES. FOR INFORMATION PLEASE
WRITE TO PREMIUM MARKETING DIVISION, NEW AMERICAN LIBRARY,
1633 BROADWAY, NEW YORK, NEW YORK 10019.

This book previously appeared in a Plume edition published by New American
Library.

 SIGNET TRADEMARK REG. U.S. PAT. OFF. AND FOREIGN COUNTRIES
REGISTERED TRADEMARK—MARCA REGISTRADA
HECHO EN CHICAGO, U.S.A.

SIGNET, SIGNET CLASSIC, MENTOR, ONYX, PLUME, MERIDIAN and NAL BOOKS
are published by New American Library,
1633 Broadway, New York, New York 10019

First Signet Printing, December, 1986

1 2 3 4 5 6 7 8 9

PRINTED IN THE UNITED STATES OF AMERICA

CONTENTS

CONTENTS

Contents

Contents

CONTENTS

Contents

Contents

FOREWORD

In 1976 The Southland Corporation's 7-Eleven Stores faced a serious problem. Armed robbers were hitting our 24-hour convenience stores with alarming regularity. We were losing thousands of dollars in stolen cash; and, worst of all, several of our people had been seriously injured during the holdups. 7-Eleven is a mom and pop operation whose employees and franchisees usually come from the neighborhoods in which the stores are located. We were concerned for the safety of these neighborhoods, as well as for our own people. We needed help.

Although the local police departments were doing everything possible, they were having to cope with a tremendous increase in crime. The local authorities simply could not put a police car in front of each one of our more than 5,000 7-Eleven stores.

We finally got the help we needed—from an unusual source. One of our senior executives had recently been involved in a robbery prevention experiment in California that used the advice of a team of ex-convicts who had once been armed robbers. The experiment had proved successful, and the California stores had experienced a reduction in armed robberies.

One of these ex-convicts turned out to be Ray Johnson. Working with our operations people in California, Ray had developed a special checklist that evaluated each store as a potential target, as seen through the eyes of an armed robber. Using this checklist, they then carefully surveyed each store, and, according to the results, made adjustments to decrease its attractiveness.

Southland decided to test Ray's ideas on a nationwide basis. The results have been almost unbelievable. In the past 7 years, armed robberies of 7-Eleven stores have *declined* by 62 percent. During this same period, according to the FBI, crimes in this category have *increased* by 50 percent in the United States.

In addition to the store safety survey, Ray's ideas on cash handling, employee training, and communication of our antirobbery program to the criminals who might hit our stores have also made an impact on the problem, the most important result being the increased safety factor for our store employees and franchisees.

Today, our company is fortunate to have an outstanding security department, staffed with top professionals who concentrate on our robbery prevention program. These people, working with Ray, have proved a formidable combination against the criminals who prey on small retail stores.

Why did Ray's ideas work? They worked because people listened to him and took his advice. He has the credibility necessary to make people act. Ray knows the inside of a criminal's mind because he was one for a long, long time. He brought his uncommon knowledge to 7-Eleven. It worked for us. If any law-abiding citizen will take Ray's advice, as we did, his or her life will be safer. At 7-Eleven we are happy to say, ''Oh Thank Heaven for Ray Johnson.''

John P. Thompson, Chairman
The Southland Corporation

PARDON

RAYMOND DEFOREST JOHNSON

Raymond Deforest Johnson has complied with Sections 4852.01 to 4852.2, inclusive, of the California Penal Code, which provide a procedure whereby persons may, after completion of their sentences, seek restoration of the rights of citizenship, and apply for a pardon. By an order dated February 7, 1977, the Superior Court of the State of California, in and for the County of San Diego, found that since his discharge from parole, he has lived an honest and upright life, exhibited good moral character, and conducted himself as a law-abiding citizen. Based on the foregoing, the court formally recommended to this office that Raymond Deforest Johnson be granted a pardon.

Pursuant to Article V, Section 8 of the State Constitution, his case was referred to the California Supreme Court for its written recommendation, and on September 2, 1980, the Chief Justice advised this office that the Court recommended that a pardon be granted.

By the laws of this state it is proper that I, as Governor, give testimony that, by successful completion of his sentence and subsequent good conduct in the community of his residence since discharge, Raymond Deforest Johnson has paid his debt to society and earned a full and unconditional pardon.

THEREFORE, I, Edmund G. Brown Jr., Governor of the State of California, by virtue of the power and authority vested in me by the Constitution and statutes of the State of California, do hereby grant to Raymond Deforest Johnson a pardon for the above offenses, except, pursuant to Penal Code Sections 4852.17 and 12021, his right to own, possess or keep any type of firearm capable of being concealed upon the person is not restored due to the fact that he was convicted of a felony involving the use of a dangerous weapon.

IN WITNESS WHEREOF, I have hereunto set my hand and caused the Great Seal of the State of California to be affixed this twenty-third day of December, A.D. nineteen hundred and eighty-two.

Governor of California

ATTEST:

Secretary of State

By

Deputy Secretary of State

INTRODUCTION

This book is about crime and how you can protect yourself, your family, and your property from it. Whether you're a homemaker alone in the house all day, a salesperson driving long distances on the job, a nurse or restaurant worker coming home at 2 A.M., a senior citizen out for a stroll, or even a kid walking to school, you're a potential victim. In America today crime casts a shadow across all our lives, regardless of our age, race, sex, or economic status. In a recent ABC News poll of 2,500 people, 85% perceived crime as being on the increase; 50% feared being victimized; 33% worried about physical injury; 25% worried about murder; and 84% of the women surveyed felt an increased risk of being raped.

Like most Americans, you're probably worried about crime—and you should be. Despite a recent 3% drop in reported crime index offenses, burglary, rape, armed robbery, aggravated assault—the cimes *you* are most likely to be a victim of—are still close to their all-time highs and are becoming more violent. You've noticed, too, that no matter how hard police and other enforcement agencies try, their efforts alone are not enough to make your city or neighborhood a safer place. At this point, you're looking for solid answers. You don't want theories or wishful thinking. You want to know just one thing: how to keep crime from happening to you.

I've been involved with crime for something like 45 years, the first 30 as a criminal and the past 15 as a crime fighter. The advice and tips in my book come from both perspectives.

Before I was released from prison in 1968 for the last time, I'd devoted most of my life to figuring out ways to separate people like you from their money. I stole my first car at 11, committed my first burglary a few years later, and graduated to armed robbery by 17. In the course of my criminal career, I spent a total of 25 years in prison, places like San Quentin and Folsom—universities of crime—where I learned the trade secrets of every kind of crook there is, from rapist to Mafia don. Of course, I'd already acquired a lot of know-how on my own, but during those prison years I improved on it. When con men recited their catalogs of scams and burglars revealed their house-breaking techniques, I listened. By the time I was led, manacled and in leg irons, through the gates of Folsom Prison, faced with two consecutive life sentences for armed robbery and escape, I had a Master's degree in crime.

During the next 15 years, five of them in solitary confinement for the only successful escape in the history of Folsom Prison, I was finally able to turn my head around. And when I got out, I was ready to put my expertise to work *for* society. Good friends went to bat for me, and I landed a job that eventually led to my becoming a research associate of an institute that specialized in the study of crime-related social problems.

One of our clients was The Southland Corporation, which came to us for advice when a rash of armed robberies broke out among its 7-Eleven stores. I knew plenty of people who had robbed 7-Eleven in the past, so I put together a panel of those guys who were out and going straight. Based on my work with that project, the Southland people hired me as their crime-prevention consultant, and the number of armed robberies of their stores dropped dramatically.

In addition to helping Southland solve their crime problems, I've traveled all over the U.S. in an effort to help other people solve theirs. Over the past seven years, I've spoken to countless groups and organizations, appeared on hundreds of radio and TV talk shows, and answered thousands of questions from people like you who want to know how to protect themselves and their families. If you watch the "Tonight Show," you've probably seen me at least one of the 38 times I've appeared as Johnny's guest. There you heard as much about crime prevention as I could give you in a 15-minute segment. In this book, you get it all: the facts, the insights, and, most of all, the strategies you'll need to prevent crime from happening to you.

My objective is to give you some very specific, practical, easy-to-follow suggestions for dealing with a wide range of criminal activity. Whatever the particular crime, there are two basic steps common to all attempts at crime prevention. First, you must *change your way of thinking*, and, second, you must *harden the target*—that is, improve you security arrangements in order to discourage the criminal from picking *you, your house*, or *your car* to prey upon.

In order to *change your way of thinking*, you'll have to take a good, hard look at the world the way it *really* is—not as you wish it were. You'll have to assess your personal risk in the light of today's crime statistics—not those of 10 or 15 years ago. And you'll have to stop doing some of the things you once did without a second thought—like throwing open the door whenever the doorbell rings. You can't continue fooling yourself with a false sense of security and saying, "It isn't going to happen to me." The odds that it won't are dropping every day.

As for *hardening the target*, if you follow my suggestions, the guy who's looking at you as a potential victim will look someplace else, where it's easier. If your neighbors do the same thing, he'll go to another neighborhood. And if everybody in the country does it, he'll turn to hot checks, go on welfare, or collect aluminum cans. He might even get a job! The point is, if each one of us does everything we can to prevent crime from happening to us personally, we'll be making the enterprise of crime much less rewarding. That's how we can turn the odds in our favor.

Remember, the time to deal with crime is *before* it happens. Although I've included information on what to do in the event you're faced with certain crimes, the point I stress again and again is that if you have to deal with crime after it's happened, you've waited too long. We're talking prevention. That's the smart way to deal with crime.

So, here it is—crime from A to Z and what *you* can do about it.

HOW TO USE THIS BOOK

Ray Johnson's Total Security is organized in alphabetical order to save you time in locating the information you need. Each entry falls into one of two basic categories:

1. Personal crime—theft, burglary, robbery, rape, etc.
2. General information on crime prevention—alarms, locks, doors, apartment, and residential security.

One glance at the Contents will direct you to information you're interested in. There you'll find subheadings that break down the entries further.

Under each main entry (as well as scattered throughout the text) you'll find related entries referenced with the words "See also." Since the book is an A-Z source, there is some repetition of information under these headings. But repetition is preferable to constant references to other pages in the text. After all, if you want information about putting an antitheft device on your car, you won't want to wade through pages of the latest information on home burglar alarms and vice versa. All you have to do is look up the crime or subject or one close to the one you seek answers to. Then if you want additional information, go to the "See also" entries and read them too.

A

Airport Crime
(See also Alarms, Auto; Car Stripping; Car Theft; Vacation Travel Tips.)

When you're getting ready for a trip, especially a vacation, crime is probably the furthest thing from your mind. But if you're one of millions of Americans who have air travel anxiety, the thought of flying probably does things to you that make you careless, forgetful, preoccupied, and an easy target for pickpockets, luggage thieves, and other criminals who follow you to the airport.

Airports are among the most crime-ridden places in town. They draw a lot of travelers, not all of whom come to jet off on the Concorde or join their group tour to Italy. Pickpockets, luggage thieves, and other criminals are mingling with the crowds at airports, and they are there to take advantage of unwary passengers.

It's natural to become excited at the prospect of taking a trip, but don't become careless and make yourself an easy target for airport criminals. By keeping your wits about you, you can have an enjoyable trip and keep yourself, and your belongings, safe from crime.

AIRPORT PARKING LOTS

If you don't *have* to use the airport parking lot, don't. Leaving your car in one of them, even for a short time, can be asking for trouble. Listen to what happened to a couple of guys I was in the joint with once.

These guys decided a branch bank near a crowded airport terminal would make a good return on a small investment in armed robbery. First, they stole a brand-new Olds (some generous soul had left the key in the ignition). Then, they tooled on out to the airport, went into the bank, grabbed the money, and quickly got lost in the crowd. Everything went perfectly.

They made it back to the car, hopped in, put the key in the ignition, and hit the starter. Nothing happened. Not so much as a grunt came from the motor. When one of the guys looked under the hood, what he saw got him hopping mad. Some dirty so-and-so had stolen the battery—and, as he later complained, in broad daylight.

Losing a battery or a set of hub caps is one thing. Losing a car you haven't even finished paying for is another. Remember, car theft from airport parking lots is on the increase.

What's the best way to avoid having your car stolen? Don't take your car. Take a bus, or a train, or a taxi to the airport. Or, let a friend or family member drive you. If none of these alternatives is possible and you must take your car to the airport, here are a few tips for airport parking safety.

1. Take the old clunker. If you're like most people who fly, you probably have more than one car. Don't be foolish and take the Stingray or Cimarron, or even your new Toyota Celica, only to leave it in the airport parking lot. It might not be there when you return.

2. Lock your car and take your parking voucher with

you. Many professional car thieves simply pay the parking fees and drive the cars out of the lot. Twenty bucks for a late model Mercedes isn't bad—especially when it buys a guy plenty of time to get the car out of the country or dismantle it for parts *before* the theft is discovered. The idea is to keep your voucher with you.

3. Don't leave things of value in the car. Especially don't leave valuables where they can be seen from outside the car. And that includes CB's and expensive digital stereos. You may think of them as permanent fixtures in your car, but only a couple of screws hold them in place and a thief probably knows just where those screws are. A car with expensive accessories is a sitting duck in an airport parking lot.

4. Protect your car and its contents with a good, reliable alarm. Keep the alarm armed (i.e., turned on), even when your car is home, safe in its garage.

5. Disable your car temporarily. There are several ways to do this, and a mechanic or even a mechanically inclined friend can show you how. Methods vary with different-make cars. Whenever I leave my car for any length of time, I remove the distributor rotor and take it with me. (I stole a lot of cars in my time, but I never once thought to carry a distributor rotor in my pocket. Car thieves don't do tune-ups. If a car wouldn't start, I simply tried the next one. Usually, I didn't have to go far.)

Personal Safety

Violent crimes in airport parking lots are on the increase—particularly late at night, in the long-term lots of big-city international airports. So, if you're going to be leaving or picking up your car from an airport parking lot, here are a few tips for your personal safety.

1. Drop your bags at the terminal before you park your

car. You can leave your bags with a skycap or with someone you're traveling with. He travels fastest who travels alone. And you can run faster without luggage, too.

2. Find a parking spot as close as possible to the following:

- The lot exit gate (cashier, security, etc.)

- The terminal bus pickup point

- The brightest, best-lit space in the lot

- All of the above

Even the bank robbers I told you about wouldn't have lost their battery if they'd thought to park nearer security.

3. Watch for anyone loitering, or sitting in a parked car, or acting suspiciously. Most people who park at airports are in a hurry to go somewhere. Those who hang out at parking lots are probably up to no good. Before getting out of your car, look around. If you see anyone loitering, don't park. Drive to the cashier at the exit and request a call for security to investigate. Do this even if you're late for a flight. It's better to miss a plane than to take a chance on getting robbed, injured, or worse. And you may save some less observant soul from becoming a victim.

4. Take a cab to your car. Upon your return, especially if you have a late night arrival, don't go wandering alone through acres of parked cars. Take a cab from the terminal to your parked car. Give the driver an extra tip to see you safely to the exit. So what if it adds an extra five bucks to your trip expenses? That's cheap when you consider the alternatives.

AIRPORT THEFT AND BAGGAGE RIP-OFFS

In prison, you have a lot of time on your hands. And one of the things prisoners do to pass that time is swap "war stories," tall tales of their exploits, much as other professionals do when they get together. A story I heard over and over again, particularly from the younger guys, was about all the good stuff they stole at airports.

> The procedure was always the same. They'd look for a baggage carousel where security was lax or nonexistent (maybe because an arrival was moved at the last moment to a different gate). Then, they'd hang back with the arriving passengers until the baggage was unloaded. If a bag they liked went around more than twice, they'd take it off and leave it on the floor behind them, as if to clear the carousel. If no one showed up to claim it after a few more minutes, they'd pick it up and leave.

The easiest marks at airports are the people who pile their stuff up somewhere and go off to buy a paper or a drink, or people who put down their cameras or airline tickets and just walk off. Sounds incredible, but airports are filled with people who unknowingly invite crime. Here are some tips to avoid becoming a target of airport crime:

1. **Leave the expensive luggage home.** As with the good car, so with the good luggage—leave it home. Bags that identify you as being well-heeled are sure to put a gleam in the eye and a spring in the step of guys like my old prison friends. Expensive luggage belongs in the back of a Rolls, under a chauffeur's watchful eye, not on an airport carousel. So drag out the battered bag you've had for years—something stout and scruffy and lockable.

2. **Don't carry more bags than you absolutely have to.** Be ruthless. Try to restrict yourself to a carry-on flight bag and a clothing bag you can hang in the cabin closet.

3. **Avoid checking bags through.** Airlines require that you attach an identification (ID) tag with your name and

address to each piece of luggage you check through. Although most airline personnel are honest, the risk is still there. Anyone who spots your address on a bag that's on its way to Nassau will know just where to go to load up the van. Or they can sell the information: A list of affluent absentees commands a good price in some circles. So think it over before you pack the closet into your 12-piece set of matching luggage. If you must check luggage through, write your office address or destination hotel or resort address on the identification tags.

4. Mark every piece of luggage in a way that's recognizable at a distance. As more and more of us travel by air, our luggage tends to look more and more alike. Even honest people can unwittingly pick up the wrong briefcase. (This can give you quite a shock when you've just settled into your seat for a 3,000-mile flight to an important sales meeting.) Mark everything you carry—suitcase, briefcase, attaché case, and especially clothing bags and the new soft luggage that's so popular—in some distinctive way. You might buy a wild shade of ribbon and tie, or sew it, to your bags. Or, you could paint X's or designs on the sides in glow colors. Another idea is to use bright-colored strips of plastic tape. Whatever you do, forget about those engraved tags that flash your name and home address for all the world to see.

CROOKED CABS AND PHONY LIMOS

If you fly a lot, you've probably heard enough horror stories to alert you to airport hustling schemes. Unfortunately, few airports of any size, anywhere in the world, are free of these hustlers. They are usually driving some excuse for a cab that would send a junkyard dog howling. But not always. Some of the classier drivers deck themselves out in chauffeur suits and drive stretched Lincolns or Caddies. Here are two ways to beat them:

1. **Don't go storming out of the terminal and into the first open car door you see.** If you're arriving somewhere new, look for the terminal information desk and ask about transportation to your destination. If the airport has a supervised cab-loading area for licensed cabs, use it; it's there for your protection. Ask a skycap or reservations attendant the going rate for a cab ride from the airport to your destination. If you get taken the long way, ask for a fare receipt and make a note of the cab number and driver's ID for future action.

2. **Avoid all solicitations from "cabbies" who have "a friend" in the front seat.** Most cities have regulations against cab drivers bringing along their friends, and a majority of cab companies prohibit it, for insurance reasons. What's at stake is your personal safety. If it turns out to be two against one, you could be a victim of robbery, rape, or assault. So even if you're late for an important appointment, take the recommended transportation—licensed cab, bus, or train—to your destination.

Alarms, Auto
(See also Car Stripping; Car Theft; Vacation Travel Tips.)

NEW CAR ACCESSORY ALARMS

It sounds like the safest way to go. But is it?

The new-car salesman delivers the pitch for the new, state-of-the-art accessory factory alarm, designed specifically for your inspired choice on the showroom floor. The $200 tab disappears into the Contract of Sale.

Curious about the details of the alarm system, you

drop by your friendly dealer's parts department and order a shop manual for it. While you're there, you see a couple of slick-looking dudes also ordering copies of the manual.

Then, one night, after the first double-header of the season, you and a friend arrive back at the stadium parking lot to find your parking space empty. One of those shady characters has caught up with you, gone through your car locks, disconnected the wiring at just the right spot, and driven off in your shiny new rocket.

This sad little story sums up the case against new-car factory alarms. Factory manufactured alarms, installed by factory trained technicians according to factory specs in thousands of similar cars, all have the same design. Control units and alarms are in the same location in each installation, the wiring is similarly routed, and switches and connectors are located at the same points. The shop manuals for these alarms usually run under $10 and are available to anyone willing to pay for them. All a thief has to do is pick up a copy of the shop manual, look for a particular make of car, and use the information to disconnect or bypass the alarm.

NONSTANDARD ALARMS

There's a better way to go. Auto alarms are available off the shelf from most auto parts stores. These packaged systems normally come with installation instructions that include a list of the tools you'll need. If you're a do-it-yourself type, you probably already have the tools. By purchasing a good alarm and installing it yourself, you'll be able to protect your car with the latest hi-tech system for a hundred bucks or less (depending on the sophistication of the alarm), a price you'd otherwise pay for the labor alone.

If you're just not the do-it-yourself kind, you'll still do

well to have a nonstandard system custom installed by your local auto accessories or electronics sales and service outlet. It won't cost any more than the factory rig, and you can tell the guy who installs it to include some surprises for any dude who has plans for your car.

SHOPPING FOR AUTO ALARMS

Whether you decide to take a shot at installing the alarm yourself or to pay someone to do it for you, you need to find out what's available. Here are some suggestions:

1. **Watch for newspaper and magazine ads for car alarms.** Clip the ads and study them. Compare features, performance claims, and prices.

2. **Comparison shop.** While you're in the shop, pick up sales literature or brochures. Ask salespeople for information and advice.

3. **Call your local auto club office.** Ask for references to reputable shops that install alarms.

4. **Consult Consumer Reports on car alarms.** Study the quality and performance evaluations before you commit to a specific make of equipment.

5. **Discuss car alarms with your friends.** You'll be amazed to find out what some of them will know about car alarms.

6. **Contact your police crime prevention office.** Call, or better yet, stop by the police station and ask for information on car alarms—including any local laws or ordinances that might affect your purchase.

HOW MUCH WILL IT COST?

The list of manufacturers of auto alarm components is a long one. Because of the increasingly high price tags on new cars, new and more sophisticated alarms are coming onto the market every day. Rather than give you less-than-timely price information, here are some general tips about price and quality.

What price should you pay for your car alarm? That depends on how much protection you need, how much car theft is going on in your area, where the car is to be used and garaged, whether you use the car for business or pleasure, how valuable the car's contents are, the market value of the car itself, its popularity, age, uniqueness and/or rarity, and on and on.

Generally speaking, you probably don't need to spend several hundred dollars on an alarm to protect a car worth $2,000. On the other hand, you wouldn't install a $24.95 basic siren burglar alarm kit in your $40,000 Ferrari. In other words, the more valuable your car, the more sophisticated, and costly, your alarm should be.

PASSIVE VERSUS ACTIVE

The terms "passive" and "active" are simply buzz words used to distinguish car alarms. Once installed, a passive alarm remains on, or armed, continuously without your having to turn it off and on every time you get into and out of your car. An active alarm *does* have to be turned off and on by the driver.

Originally, all car alarms were active. But drivers tired of having to arm and disarm them. It was too easy, and convenient, to forget. So the electronics engineers got busy and designed an alarm we wouldn't be able to forget, one that would protect the car all the time, whether or not we turned it on.

Although there are several types of passive systems, most are turned off when you insert the key in the ignition and are turned on when you remove the key. An automatic delay built into the system permits you and your passengers enough time to get into and out of the car. As in nearly all car alarm systems, accessory pin switches installed in the hood and deck lid will also trigger the alarm.

Automatic Passive Alarms

If you're in the habit of leaving your keys in the ignition, there's help for you. One alarm model on the market automatically arms itself when the engine is turned off and can only be disarmed by entering a four-digit code into a control module mounted inside the car. Even if you leave your keys in the ignition, the alarm will be triggered by anyone entering the car who attempts to start the engine without deactivating the alarm by punching in the digital code. The engine, of course, will not start unless the code is entered.

Keyless Active Alarms

If you're the disciplined type, you can buy a top quality active alarm system, usually for a little less money than a comparable quality passive system. There are two principal types of active car alarms: the keyless model and the key-operated model.

The keyless model works as follows: When your car door is opened, a door switch turns the courtesy light on, causing a slight drop in battery voltage. The alarm control unit senses the voltage drop and sets the alarm to go off after a short delay period, usually 15 to 30 seconds (some are adjustable to longer periods). Hidden somewhere in the car, under the dash or under the seat, is the control unit with a manual switch. To disarm the alarm, the switch must be turned *off* before the delay period elapses. Conversely, the switch must be moved to the *on* position whenever you leave your car. If you forget, the car is left unprotected. And if you're not particularly ingenious about where you hid the control unit, the turkey who gets into your car will find it and turn it off.

Another keyless system is the infrared activated system,

which uses a wireless remote transmitter the size of a pack of cigarettes. You use it the same way you use your remote control for your TV. With it you can turn your car alarm on or off from outside the car. This avoids delay and reduces the opportunity for the thief to get inside your car and turn the alarm off.

Key-operated Active Alarms

The key-operated active systems are the oldest and most familiar alarms. They're the simplest to operate, although not necessarily easiest to install. They are also the most economical. The better quality key-operated alarms are probably the most trouble-free of all car alarms. With these alarms, after you lock your car you put a key in a switch located somewhere on the outside of your car to turn the alarm *on*. Before getting back into your car, you insert the key in the exterior switch and turn it *off*.

PAGING ALARMS

Paging alarms use citizens' band (CB) transmitters installed under the seat or dash of your car to warn you of any tampering. The transmitter is connected to a network of highly sensitive sensors installed in the body of your car that "listen" for sounds of tools or car movement. In the basic model, no alarm sounds in the car itself. You receive the alarm signal through a pocket receiver, or pager, which emits an audible beep. The theory is that you'll be able to call the police in time to catch the car thief in the act. The transmitter receiver range is up to 2 miles.

If you're more interested in scaring the thief off than in adding to the prison population, an optional vehicle light/horn system will honk the horn and blink the lights at the same time the transmitter is beeping your page.

ALARMS THAT STOP THE ENGINE

It's not unusual for a determined professional car thief to find a way to disconnect a screaming car alarm and drive off. For this reason, many better alarm systems include an additional feature called the *ignition cutoff switch*. As described in the passive system, when an alarm with this feature is triggered by an intruder, it's impossible for him to start the engine and drive off. Not even hot wiring or bypassing the ignition is possible unless the thief can locate the ignition cutoff switch, which is usually a more complicated operation than the crook has time for. To make things even more interesting, some models permit the engine to start and run 20 or 30 seconds before killing and then prevent it from starting again until disarmed.

Ignition cutoff devices are a standard feature of a number of alarms that cost about $50 and up. They can also be purchased as accessory units for as little as $14.95.

A NO-COST TIP

In my work with crime prevention, I've often found the best prevention to be simple common sense things that don't cost a dime. Here's an example.

I've got a 1964 Lincoln I drive around that's still in pretty good shape. A while back, I decided to upgrade some equipment on my car. When I came up with an extra knob on the dash that wasn't hooked up to anything, I ran a couple of wires from it to the ignition. To start the car now, you not only have to turn the ignition switch on, you also have to pick out this knob from about five others that look just like it.

SHOPPING AND INSTALLATION TIPS

Shopping for just the right alarm or do-it-yourself components can be tricky. Here are tips that will take some of the guesswork out of it:

1. Don't assume all alarms are legal in all places. Before buying alarms or components for do-it-yourself installation, discuss your plans with the local police to make sure they're legal where you live. For example, sirens in alarms are illegal in California. If you plan to use your car in California, install a horn and light alarm instead of a siren.

2. Some mechanical or electrical features may affect your choice of alarm. For example, if your car is a classic or collector's item, it may have a 6-volt electrical system, rather than the 12-volt system most manufacturers converted to in the 1950s. And, although most U.S. cars traditionally have had negative ground systems, some foreign makes have positive earth (ground) systems. Read your owner's manual and the instructions that come with the alarm carefully before you install it. Improper installation may damage the alarm and void its conditional warranty—and may damage your car as well.

3. In choosing an alarm and its components, consider how the car will be used—whether for business or pleasure—and who its passengers will be. For example, don't waste your time and money on an alarm triggered by sensors if you're constantly leaving kids or pets in the car while you're in the bank or shopping. An expensive passive system limiting entry and exit times isn't for you if the car is used regularly for handicapped or elderly people. Discuss the way the car is to be used with sales people and others you talk with before purchasing an alarm.

4. Once you have decided on the type of alarm best suited to your needs, look for one that offers as many of the following design features as is practical and useful:

- Unpickable, round-key security lock switch

- Wireless components

- Passive activation

- Motion/vibration sensors to detect jacking or towing

- Switches or sensors to protect doors, hood, and trunk

- Separate circuits to protect stereo and CB

- Ignition cutoff

- Starter switch cutoff

- Automatic fuel cutoff (gas or diesel, except Mazda and Mercedes)

5. If you buy your system installed, make sure you and the guy who's doing the work understand each other. Get a written estimate, which includes not just the price, but also the make of alarm and the work he's charging you for. Get at least three estimates before deciding who to have do the work.

6. If possible, be on hand when the installation work is done. And don't accept work as complete until it fulfills the written agreement.

7. Make sure you get the alarm and installation you're paying for. Even if you buy a system off the shelf, ask for the box or carton the system was shipped in from the manufacturer, together with all instructions, specifications, and warranty.

8. Once your alarm is in and working, advise your insurance agent. You may be entitled to a lower insurance rate.

Alarms, Home and Business
(See also Alarms, Auto; Burglary; House Security.)

Moats and alligators—that's the way I think of the very sophisticated and expensive security systems rich people put into their mansions and estates. These invariably include carefully engineered and designed alarm systems connected by phone company lines to 24-hour control centers in private security companies. They're sometimes called *central station alarms*.

I have a good friend of many years who, until he was finally caught, was a preeminent West Coast burglar. He's now retired and living a sedate life. At the height of his career he specialized in celebrity homes—movie stars, the rich and famous, nothing but the highest-priced houses. He was so successful at eluding discovery that they dubbed him the Bel Air Phantom.

Partly due to my friend's success, the time came when an aggressive security service company introduced its new moats and alligators alarm system to the affluent citizens of communities such as Bel Air, Westwood, Brentwood Heights, and Beverly Hills. The company did a hot-shot marketing job, and in no time at all everyone who was anyone had one of those moats and alligators setups around his or her castle.

Suddenly, my friend found his livelihood threatened by this new technology. What to do? After giving the problem some thought, he rented a big house in Bel Air, then called the security company and had one of their new alarm systems installed. He studied the thing inside and out, found out what made it tick, and, more importantly, what would make it *stop* ticking. He memorized the sequence and placement of control box, switches, circuits, and other devices. When he was sure he could defeat it, he was ready.

He didn't have any trouble identifying the houses with the new alarm systems. The security company had very conveniently placed ads in the form of official-looking gold medallions imprinted with its name in big black letters on the front gates, pillars, and fences surrounding these treasure domes. The signs were of course meant to impress on the neighbors the fact that they, too, needed one of these alarm systems, as well as to inform would-be burglars of the futility of lusting after a house equipped with one of these impregnable alarms.

The warning certainly sold a bunch of high-priced alarms for the security company. But all it did for its customers was tell my friend that the people who lived in those houses had a lot of interesting stuff they wanted to protect and that they were protecting it with an alarm system he knew better than his girlfriend. He just drove around and picked them off, one by one.

The moral of the story is twofold:

1. When you start thinking about how to install your own alarm system or about having one installed for you, make sure it has many odd twists unique to your particular system.

2. Do *not* avail yourself of the installer's or manufacturer's free medallions or warning stickers to be attached to doors, windows, etc. Don't post warnings of any kind telling intruders what or who is protecting your house. Few of the crooks I've known were impressed with those stickers. If they do anything, it is to make you more interesting. Protection is what the *alarm* is for.

BUY WHAT YOU NEED

If you're attracting burglars with the expertise of a Bel Air Phantom, you'd better have the best that money can buy. A moats and alligators system may be right for you. On the other hand, if the only thieves you need to worry about are the amateurs common in suburbia, all you'll need is an alarm that makes a lot of noise during daylight hours and turns on a lot of lights at night. These are relatively simple, easy-to-install, and economical alarm systems that will alert you to anyone who tries to get in while you're home and will drive anyone off when you're not.

The fact is, most of us don't need moats and alligators. The stuff we have just isn't that interesting. Whether we live in a crime-ridden city or in suburbia where juvenile crime is the problem, noise and light—the two least expensive and most accessible forms of alarm—will prevent 98 out of 100 potential intrusions into our homes and apartments. You may think differently, but I've got more than 30 years of experience that says no one with over three points of IQ is going to stand around, day or night, with all the bells, whistles, and blinking lights of a modestly priced home alarm installation going off, waiting for someone to come ask him what he's doing there. He's going to pass right on down the street and try some house that's easier to get into.

Security Survey

Give yourself the following security survey to determine just how much alarm you really need:

1. Do police reports reflect a high rate of burglary and other property crime in your community?

2. Does your home or apartment have ground floor windows and entries, connected garage, roofs or fire escapes, climbable trees, etc., which make it particularly vulnerable to intrusion?

3. Have your immediate neighbors experienced burglaries, attempted burglaries, or intrusions?

4. Do you own an expensive home or have expensive cars that cry out *wealth* to passersby?

5. Do you own a lot of valuable things such as paintings and antiques, collectibles, jewelry, gold and silver, stereo and video equipment, computers, cameras and photographic equipment, guns, power tools, etc?

6. Does the nature of your work make you vulnerable? For example, do you have diplomatic or political connections? Do you control large sums of money? Do you store valuable drugs or expensive equipment in your home or office?

Even if you answered all of the above "no," you should have some sort of alarm. Remember, an alarm system also protects your life and the lives of your family members. If you answered "yes" to more than one or two questions, you should have a good noise-and-light alarm.

If you've answered "yes" to more than three questions, you're probably among the small group of people who really *need* moats and alligators protection.

Ask for Advice

Professional advice from people in the alarm business and others who can help is as near as your telephone. Look under "Burglar Alarm Systems—Dlrs. Installation & Svce." in your local *Yellow Pages*. Talk to your insurance agents or brokers. Ask your police crime prevention units for their advice and ask them to recommend the best service in your area.

1. Home insurers. Talk to your insurance broker or agent. Some insurance companies have home security consultants well briefed on local crime. They'll visit your home and conduct a free security survey of your property. A few companies even give discounts to policyholders with security systems.

2. Your local law enforcement crime prevention unit. This is usually a special unit or squad within your police precinct or sheriff's office that has the responsibility of keeping you and your neighbors alert to local crime. In most cases, they are members of the regular patrol who have the added responsibility of crime prevention work with local citizens groups.

Keep in mind that some crime prevention units, particularly those in higher-income suburban areas in which there are the most home burglar alarms become disenchanted with noise and light alarms. This is because of the high incidence of false alarms resulting from homeowners' poor choice of systems for a given set of conditions and careless installation jobs. To take some of the pressure of responding to false alarms off the local police, they may try to steer you to the expensive private alarm systems companies that offer central station burglar alarms.

3. Neighbors and local businessmen. Ask your neighbors and the people you work with or do business with what alarm systems they've had experience with and what they recommend.

ULTRASONIC AND PERIMETER ALARMS

Although there are many varieties of burglar alarms, there are just two basic kinds of alarm: ultrasonic (motion detector) and perimeter.

1. Ultrasonic alarms. These are supersensitive devices designed to detect movement in sound waves in the air around them. The models available to us are suitable only for single room protection. Obviously, the major drawback is that the intruder has to be in the house to set off the alarm. (Don't let anyone tell you that you can use them out of doors because you can't. Birds, animals, the wind, literally any movement can set them off.)

2. Perimeter alarms. These are designed to detect attempts to open doors or windows, break glass, move safes, warn you of intrusions, and generally protect the perimeters of a given area—usually a building. Perimeter systems of varying sophistication provide noise and light and are the basis for most professional alarm systems, including the central station alarms.

YOUR CHOICE OF ALARM

In the seven years that I've been working as a consultant with the 7-Eleven stores, I've been asked by the makers of many alarms and protective devices to endorse their wares. I've only endorsed a few, because there's a lot of junk out there. My advice to you is to take your time about making a decision. Read everything you can find on the subject of home burglar alarms. Make a collection of manufacturers' brochures and literature. Get as much advice as you can.

Here are a few of the most common choices available to most of you with average security needs and budgets.

1. Door and window electronic noise alarm. If you've got a house full of kids and the budget for extras just isn't there, these little alarms are easy on the pocketbook and easy to install. Models start as low as $5; the key and keyless digital models are slightly higher (from $10 to $20 each). You need one alarm for each door or window you want to make secure. For most installations, all you'll need is a screwdriver (see Figure 1).

2. Ultrasonic motion detector alarms. Ultrasonic alarms are available in a broad range of models and prices. Room protecting alarms you simply plug in and aim range from

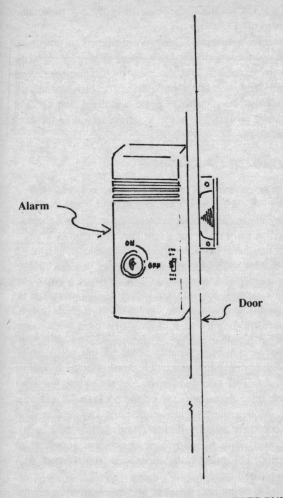

Figure 1. DOOR AND WINDOW ELECTRONIC NOISE ALARM.

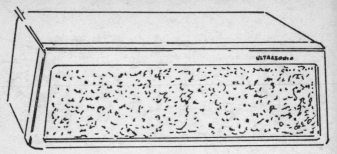

Figure 2. ULTRASONIC MOTION DETECTOR ALARM.

$40. If you rent or lease your apartment or office or store and perimeter alarm installations are prohibited by your lease, an ultrasonic alarm is for you (see Figure 2). Its advantage is that it requires no installation. The disadvantage is that it protects only the interior area. Unless it's used with more expensive and sophisticated perimeter systems, which do require installation, the ultrasonic alarm sounds only after the intruder is already inside your home or office.

3. Perimeter alarms. If you live in a house or own a business in which there's money or other valuable property or if you live in a suburb in which there's a lot of prowling and burgling, you need a perimeter alarm (see Figures 3 and 4). These can be simple or extremely sophisticated, wired or wireless, open or closed circuit. A simple, wired, normally closed circuit alarm, including control box, alarm siren or bell, door and window switches, and necessary wiring is available from leading electronic and hardware outlets in a box for under $150 ($100 on sale).

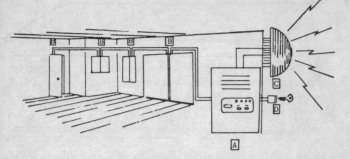

Figure 3. WIRED (NORMALLY CLOSED) PERIMETER ALARM SYSTEM.

A, alarm system control box; B, magnetic door and window switches (with wiring connections to control box); C, accessory super-loud outdoor bell or siren; D, accessory outdoor key switches can be installed at (up to 10) entrances for convenience in arming or disarming alarm system.

4. Microprocessor controlled perimeter alarms. These state-of-the-art alarm systems are recent, more sophisticated, and more expensive variations of the basic wired perimeter alarms. They operate according to one of two basic principals: (a) FM (frequency modulated) through-the-air radio signals or (b) line carriers using binary encoded signals transmitted through existing house wiring.

One advantage with this alarm is that the transmitters can be placed at doors and windows throughout the structure without the need for extensive wiring connecting them to the alarm. Only minor installation, similar to that for the much cheaper door and window alarm units, is required.

You can incorporate two independent systems in one central control unit to make what is called a *zone system*. A disadvantage of this system is the cost, which starts at $500.

Shopping for Your Alarm

Here are a few things to keep in mind when you start shopping for your alarm:

1. If you live in a rented house or apartment or if the alarm is for leased business space, talk to your super or landlord before you buy. Your lease or other contractual agreement may require that you get permission before alterations or installations can be made.

2. Watch newspaper and circular ads for a few weeks before making your final decision. Burglar alarms are hot items with hardware and electronics retailers. You may be able to pick up the hardware at sale prices.

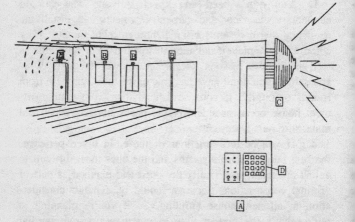

Figure 4.   WIRELESS PERIMETER ALARM SYSTEM.

A, alarm system control box receives signal from wireless door and window transmitters; B, each transmitter installed at doors and windows requires a 9-volt battery for power. C, accessory outdoor bell or siren is wired to control box. D, programmable digital on–off control means you can program your own code. Built-in delay allows you to enter and exit without setting off alarm.

3. Shop the catalogs of electronics, hardware, radio–TV, and department stores. Mail-order units of comparable quality to those available through retail outlets are often priced significantly lower.

INSTALLATION

When you've determined your chances of intrusion and decided what the most dependable alarm for your particular circumstances is, one final decision remains: whether to install the alarm yourself (if installation is necessary) or to have it installed.

1. You don't need an expert. Anyone who can do elementary electrical and carpentry projects can install the residental alarm systems offered through retail outlets. They come with complete installation instructions. For economic as well as security reasons, the preferable way to go is do-it-yourself. If you don't have the time or the tools, get help from friends or family. If you use a handyman from your apartment house or an outside worker, check his references and make sure he's honest.

2. If you're building a new home. In my experience, the best burglar alarm systems and the ones most difficult to defeat are those specifically designed and planned as part of original construction. Security should be a major consideration in all new house building. So if you're planning to build, that's the perfect time to put in a first-rate burglar alarm system.

3. Don't rush it. Whether you're doing the work or having it done, allow time to set it right. Even if you have to give it more of you time, or pay for someone else's, believe me, the better your installation the greater its value as a deterrent will be. Remember, *the least obvious systems are the best*. The less seen of your alarm, the better.

4. If you're buying a custom system with installation. Get at least three estimates for comparable hardware and installation work. Insist on written estimates, specifying just what your money will buy. If a company refuses to put it in writing, cross it off your list.

5. Good burglar alarms should be heard and not seen. In other words, don't paint your alarm box red and hang it on your rec room wall. And don't run alarm wires where they can be seen easily. Buy switches that can be concealed in window and door frames, not those attached slapdash in the most predictable and obvious places. Be sneaky.

6. Keep the location of major components—control box, bell or siren horn—secret. If possible, mount the control box in a windowless closet or utility room with a good deadbolt lock on the door. Place your superloud outdoor alarm bell or siren horn as high as possible, both to keep it out of reach of intruders and to give its sound the best possible range. A good place for alarm components in a house is in the attic, behind the ventilator opening.

7. Be a perfectionist! See to it that *all* manufacturer's recommendations are followed *to the letter*. An improperly installed and malfunctioning alarm will give you more problems than protection. In some parts of the country, they can even result in fines for violation of town nuisance ordinances.

8. Repair loose doors and windows. Do this *before* installing perimeter alarms. Otherwise they might shift or rattle from wind or other causes and set off false alarms.

9. Make certain that measurements and tolerances specified for installation of door and window switches are exact. Don't make the mistake (an all too common one) of thinking that $\frac{1}{32}$ inch, more or less, won't make a difference in how the switch works. Keep in mind that improper installation of switches and sensors is a major cause of false alarms.

FALSE ALARMS—FACTS AND FABLES

Police estimate that 90% of their responses to residential alarms are to false alarms. How many are *truly* false is hard to say. An officer responds to a report phoned in by a neighbor that an alarm is sounding. When he arrives, the alarm is indeed sounding, there's no intruder in sight, and all the doors and windows are secure. But that doesn't mean the alarm was false. In many cases, the alarms did what they were meant to—a window was slightly jimmied or a door was cracked, setting off the alarm, and the burglar took off.

If you have an alarm that consistently gives false alarms, go back over your installation with the manufacturer's specifications in hand. Check all tolerances carefully. You may find that the problem is faulty components, but I'll lay odds you'll find the alarm was not installed properly.

As a young robber, I got wise to false alarms early in the game. I'd heard about a company that paid its employees in cash and kept a lot of money on hand. I decided to rob it. The problem was getting in when they opened the safe in the morning. After casing the place for a few days, I decided that the best time to go in was the night before payday. Unfortunately, they had the best security company alarm system available at the time. I'd just about given up when I found a window that was just loose enough so that, by banging the frame with the edge of my hand, I could set off the alarm. I banged away, then ran across the road, dived into a culvert, and crawled into a storm drain to wait.

In a few minutes a security company car pulled up, a guard got out, went into the building, reset the alarm, flashed his light around, came out, checked the windows and doors, flashed his light around some more, got into his car, and left. I waited a while, then crossed the road, went back to my window, rattled it to beat

28

hell, set off the alarm again, ran back across the road and dived into the drain again. Back came the security company car; this time two guards got out of the car. More flashlights, more crawling around the place, then off in their car again. Another hour or two, another rattle—by then it was an hour or two before sunup. Back came the security company car. This time they'd dragged the maintenance man along with them. They stayed inside a long time. When they finally came out, I could hear them cussing the alarm and agreeing the best thing to do was shut the damn thing off till they had a chance to fix it. I went in and waited. A few hours later I got $65,000 from the bookkeeper and went home.

So much for false alarms!

Apartment Security
(See also Alarms, Home and Business; Burglary; Doors; Locks; Real Estate Risks; Windows.)

About a third of Americans live in apartments, the majority of them rented apartments. If you're like me, you live in or near a major city, close to your job, business, shopping, and entertainment, and your next apartment will probably be in a similar setting.

Now's the time to give some thought to the changes that have happened in our society since you last looked at apartments. If you moved into your current apartment in 1977, violent crimes have increased almost 31% in the years since then; the increase is more like 50% since 1971. So, before getting bogged down in the real estate section, think about the local crime news. When you're planning to move, the issue of security has to be placed high up on your agenda.

The local police crime prevention officer can be helpful in this department. Ask for the names of nearby communities with the lowest crime rates. If you find an area you like but can't get information about it, stop in at a local coffee shop or other business, or try to talk with the people living there.

CHECKLISTS FOR APARTMENT HUNTERS

When you're moving from one part of the country to another, you need to be especially careful. If it's a company transfer, check with the personnel relocation manager at your new assignment or talk with local real estate people. Ask the questions on the following list. Make sure they understand that security is a top priority with you. These questions can be helpful even if you're moving to a new area in the same part of the country.

1. What kind of crime rate does the community have?

2. Is it located away from high-crime areas?

3. Is it stable—or do a number of vacant or deteriorating buildings indicate the neighborhood is getting run down?

4. Are the streets clean, well-lighted, and regularly patrolled by police?

5. Do the reputation and quality of services, such as sanitation, police, and fire department, indicate a well-run community?

6. Do the other homes and apartment buildings in the neighborhood appear well cared for?

With regard to your potential new apartment, you should consider the following:

1. Does the building itself appear well managed and maintained?

2. Is there some form of security at the front entrance— doorman, guard, closed-circuit TV, or, at the very least, a strong forced-entry resistant lock?

3. If the building has parking, is it well-lighted and secure? Is there an attendant or guard on duty or a closed-circuit TV system?

4. Is the particular apartment you're considering secure? Keep in mind that an apartment is secure if:

- It has metal doors or metal-sheathed, solid core wood doors rather than hollow core wood doors.

- The door frames are metal.

- The doors are equipped with dead-bolt locks in addition to standard door locks (double locks).

- There are secure locks or otherwise burglar-proof devices on ground-floor windows, skylights, and windows next to fire escapes.

- Elevators are well-lighted and equipped with intercoms monitored by security or doormen.

- There is good building security to prevent unauthorized people from gaining access to basement, stairwells, or roof.

5. Does the rental or asking price seem unusually low? Don't let a bargain basement number attract you to a high crime area with poor or nonexistent building security. There are such things as good deals, but nine times out of ten an unreasonably low going price signals problems. Let someone else grab the bargain.

YOUR RIGHTS AS A TENANT

After spending about 20 years behind bars, I began to realize that even prisoners are entitled to a certain degree of personal security. I studied law books from the prison library and together with other members of the

Inmate's Council, I made sure we got the security we were entitled to.

If you pay rent, find out about your landlord's legal obligations for your security. (Many municipal, county, and state court buildings maintain taxpayer-supported law libraries that are open to the public—and their librarians usually are happy to help you find what you want.)

When it comes time to write a new lease or renegotiate an old one, a knowledge of landlord responsibility for tenant security can give you a lot more clout in obtaining security concessions and improvements.

IMPROVING THE SECURITY OF YOUR BUILDING AND YOUR NEIGHBORHOOD

Two things you learn early in prison life: (1) staying alive is never easy, you've always got to work at it when you're surrounded by violent people, and (2) you get nothing in prison without community action. Today's cities are looking more and more like prisons, and those of us who live in them are more and more the prisoners of the murderers and rapists around us.

During my travels around the country in behalf of crime prevention, I meet hundreds of people who live in metro areas where violent crime is on the increase. They tell me my advice sounds good, but that they can't put locks on the bedroom door because of the baby, or because the bathroom's in the hall, or whatever. I tell them to put locks on all their doors and keep them locked. They tell me they can't live like that, that it would be like living in prison. I get the feeling they just don't want to face reality; they don't want to admit things are really the way they are; they don't *want* to change their thinking.

But we *must* change our thinking if we are to push crime rates down in today's society. In our cities, those of us who live in apartments can only do it by working at it through community action. The following is a list of things *you* can do about security in your building. Any of them you do will pay dividends in safety and in a better life for you and your neighbors.

1. Get to know your neighbors. Make it a point to meet and discuss security with other tenants on your floor, particularly those on either side of you and across the hall. It's important that all of you know who's supposed to be going in and out of that door.

2. Look out for one another. Tell your neighbors about your alarm system and let them hear it so they'll know if someone is trying to break into your apartment.

3. Be on the lookout for strangers. Be alert. When you see strangers wandering around the halls or riding on the elevators, avoid confronting them directly. Instead, notify your building security or superintendent right away.

4. Join or form a tenants' group. Apart from its usefulness as a political force in dealing with uncooperative building managements and absentee landlords, it's the most effective way to monitor building security arrangements and to give your building representation in neighborhood groups.

5. Join or form neighborhood watch groups to patrol local streets. Share patrols with other tenant groups from adjacent or nearby apartment buildings. Equip the patrols with walkie-talkies in contact with apartment command posts assigned on a rotating basis with a volunteer standing by, ready to summon police.

6. Don't enter elevators with strangers. If you don't recognize them, don't get on with them. Take the next one. If someone unfamiliar gets on after you, get off. Don't take chances.

SECURITY TIPS FOR YOUR APARTMENT

If you've achieved good neighborhood and building security, you've gone a long way. Next is your own apartment. It represents an important third line of defense. Its security shouldn't be overlooked.

While You're Out

Eighty percent of attempted burglaries take place when no one is home—usually during the day. One of the best alarms for an apartment, and the most trouble free, is one of those yappy little dogs—a Yorkie or Scottie or Shih Tzu. The would-be burglar will ring first to make sure nobody's home. If he's answered with a bunch of yips and yaps, nine times out of ten he'll pass. He'll go down the hall to another apartment where there isn't a dog or other security measure.

Supposing he does decide to brave the little dog's wrath? The next thing he'll do is try to *loid*—slide a credit card or other flexible material against your bevel door lock—or to pick the lock and go through the door. If you have a good, solid hardwood door, preferably sheathed with metal and secured with a quality dead-bolt lock, he's not going to go through the door—unless he has all day and a carful of tools (and he won't). And even if he does crack your door, chances are he won't go much farther if you've installed a good, loud alarm.

While You're In

Statistically,, break-ins while residents are home are very rare. Unfortunately, when they do happen there's a much higher probability of violence. If you hear someone trying to get into your apartment at night, don't tippytoe around and whisper or hold your breath and hope he'll go away. Sit up in bed and shout, "George, I think I hear a burglar!" Shout it even if there is no George, or Georgette, as the case may be. Avoid contact, but do make noise: Shouting, sirens, or a blaring stereo will get him out of there fast.

Here's a list of security precautions you can take. It's not intended as a program but simply as a list of things to choose from to fit your particular household and life-style.

1. Keep your doors locked. Do this all of the time, whether or not you're in your apartment—day or night.

2. Buy a wide-angle door viewer. Your neighborhood hardware store sells them. Install it or have it installed in your front door. Use it when people arrive at your door. If you don't recognize them, tell them to leave their card in the mailbox. If they claim to be delivering something, tell them to leave the delivery by the door or with the doorman.

3. Purchase aerosol fog horns or hand-held personal security alarms. Keep one on your bedside table as part of your noise-making tactics if you hear a nighttime intruder.

4. Install battery-operated electronic door and window alarms. This ensures advance warning of any attempted entry and frightens off potential intruders. Keep them armed (turned on) at all times.

5. Don't rely on security chains. Don't place too much faith in the security chain on your door. Chains are useless on most apartment doors, which are usually made of wood with wooden door frames. A husky teenager can easily push his way in, pulling your security chain right out of the door frame. Some hotels, apartments, and homes have metal door frames, and in that case, a strong, heavy duty, hardened security chain, properly installed, will provide some measure of security. But it's safest never to rely on a door chain. Don't take chances by opening your door at all. When you hear a knock or the bell, keep the door shut and locked, and check the caller through your door viewer.

6. Harden the target. Install heavy wooden or metal

doors and door frames in all entries where heightened security is necessary. Don't put a $50 lock on a $10 door.

7. Use rubber doorstops. The housewares section at the shopping center has them. Insert the rubber door stop under your locked door for additional security. This considerably strengthens even a flimsy door and few would-be intruders will anticipate this device as a possible impediment.

8. Install dead-bolt locks on all doors. Do it even if your doors are not yet *hardened*. (See "Doors.") Keep them locked at night and when you're out. This will provide a measure of protection for occupants of the rooms as well as confuse an intruder as to the rooms' use and the importance of their contents. Particularly if you're a woman living alone, a deadbolt lock on your bedroom door is a vital second line of defense. At least you won't wake up to find an intruder standing at the foot of your bed.

9. Set up a wireless intercom with your neighbor. Find a neighbor whose at-home hours are similar to yours, so that you can alert one another to emergencies or problems.

10. Protect the nursery or children's room. Keep an additional wireless intercom with call switch open beside your child's crib or bed so that he or she can call you in the night.

Armed Robbery
(See also Automatic Teller Machines; Mugging; Personal Safety; Retail Store Crime.)

Robbery was my specialty. In the years before I went back to Folsom for the last time, I hit movie houses, supermarkets, nightclubs, banks, drive-ins, loan com-

panies—you name it. I never went after people in the
street because that wasn't my style.

I waved a gun in a lot of people's faces in the course
of these stickups, but I never hurt anyone. I can tell you
this, though, I never did a job where I wasn't psyched
up, excited, and scared. And I've never met a robber
who didn't feel the same way when he pulled a job. I
was always anxious to get the money and get out, and it
never happened fast enough. But I was lucky. The
people in the places I robbed didn't give me any trouble,
and they let me get away from them, fast.

If *you* are ever the victim in a robbery, remember my
experience. Get it over with as quickly as you can, and try to
make sure that others around you do the same. Robbery isn't
a game of chicken, group therapy, or a happening. It's a
crime of deadly force, an extremely dangerous confrontation.

If I were to ask you where most robberies happen, you'd
probably answer banks. But you'd be wrong. Of the 537,000
robberies in 1982—a robbery every 59 seconds—only 6,981,
a mere 1.3%, were in banks. In fact, a bank is your safest
refuge from armed robbery. You're much more likely to be
robbed while walking on the street or driving on the highway,
where 54% of all robberies happen. Another likely place for
you to be robbed is in your own home, where over 11% of
robberies happen.

That's bad news, since all of us use our homes, streets, and
highways more often than we do banks. The good news is
that you can do things to reduce your chances of being
robbed.

HOW TO AVOID BEING A VICTIM

There are ways to reduce your chances of becoming a victim,
whether you are at home, on the street, or on the highway.

At Home

1. Never let strangers into your home. Install a door viewer and use it. If your caller's a salesman, tell him to leave a card or brochure outside your door. If you live in an apartment house and he claims to be a repairman, tell him to come back with the manager or the super. Whatever the story, don't let him in until you have proof positive that he's legit.

2. Don't invite outsiders in. If strangers drive up your driveway or hang around outside your apartment door, don't march out and ask them what they want. If they have a legitimate reason for being there, they'll let you know. If they don't, call the police and let them handle the situation.

3. Drive away from home. If you return home to find a carload of shady-looking characters parked up the street, keep driving. They *could* be your neighbor's in-laws, but the price of guessing wrong isn't worth it. It's safer to drive around for a while and come back later. If they're still there, find a telephone, call the cops, and let them investigate.

4. Walk away from home. If there's a weird-looking character hanging around the street outside your townhouse or apartment or standing in the vestibule entrance to your building, don't go in. Keep walking till the person's gone.

5. Keep a clear view. Remove the opportunity for predators to surprise you by keeping shrubbery around entrances and windows trimmed back. Well-trimmed, thorny hedges also discourage intruders. (See "House Security.")

6. Keep your home well lit. If there's a possibility you may be returning home after dark, keep the driveway and entrances to your home well lit. Use automatic timers to turn on interior lights and outside floods.

On the Street

1. Never stroll aimlessly. Know where you're going and act like it.

2. Don't overdress. Do not dress expensively just to show off. Do not wear valuable jewelry when you plan to be in the street. Attract as little attention as possible.

3. Walk on well-lighted, main boulevards. Do this whenever possible. You may have to go several blocks out of your way, but what you lose in time you gain in safety.

4. Be alert and conscious. Pay attention to what's going on around you as you walk. If you think you're being followed, head for the nearest store or open business and ask someone to call the police. Don't run into parking lots, unpopulated buildings, or lobbies. Don't try to call for help from a public telephone on the street (even those with doors are not secure from a determined assailant).

5. Get away from suspicious people. If you spot someone who looks suspicious, get away fast. Cross the street. If necessary, walk a few blocks out of your way to avoid confronting the person.

6. Watch the traffic. On city streets, keep an eye out for occupants of parked cars, slow-moving traffic near you, the spaces between parked cars, doorways, and alleys.

7. Carry protection. Carry an aerosol fog horn or electronic personal alarm with you at all times. Use it if anyone menaces you or refuses to leave you alone.

On the Highway

1. Keep your car doors locked and windows up. Don't open your car doors to anyone unless you're sure of the person's identity and purpose.

2. Never pick up hitchhikers. If you do, you're leaving yourself wide open—and you're on your own. Hitchhiking is against the law in many states, and that gets your insurance company off the hook if anything happens.

3. Call for help and then wait. If you have trouble on the highway, drive slowly—or walk, if your car won't move—

39

to the nearest emergency phone and call for help. Make sure your car always has a half-dozen safety flares in its emergency kit. Put flares out 200 feet behind your car to warn oncoming motorists, turn on your hazard lights, and raise the hood. Unless you've been able to get your car out of the traffic lane, off the pavement, and onto the shoulder of the road, it's not a good idea to stay in your car. If you have been able to get it out of the way of oncoming traffic, lock yourself inside and wait for the police or highway patrol to find you. If other motorists stop to offer aid, roll down the window an inch or so and ask them to call the police from the nearest phone.

4. **Don't stop to help motorists beside the road.** Their distress signs may be a ruse. Make a mental note of their location. At the next exit, rest area, or emergency phone, notify police or highway patrol.

5. **Lock your car at all stops.** This is true no matter for how short a time. Always check the back seat and floor of your car before getting in.

6. **Notify people.** If you're going far, tell the people you'll be visiting your route and expected time of arrival. Or tell friends or neighbors before you leave that you'll call them when you reach your destination. Advise everyone that if they don't see or hear from you within a reasonable length of time to call the police. Don't take long trips without telling someone of your travel plans.

7. **Beware of bump-and-rob attempts.** This is a tricky one. Uniform traffic laws say you must stop immediately in the event of an accident, even if you're merely a witness. Unfortunately, robbers have learned how to use this law to their advantage. In a typical bump-and-rob crime, a car will strike yours, sometimes from behind, sometimes by cutting you off. When you get out of your car with license in hand, you're confronted by a gunman who relieves you of cash and

valuables, takes your ignition key or otherwise disables your car, and drives off. Your only defense is to guess what he's up to and avoid stopping or being stopped until you can notify the police. Be very suspicious of anyone who strikes you from behind or cuts you off on the highway.

7 SURVIVAL STRATEGIES FROM AN EX-ROBBER

Armed robbers are made, not born. Very seldom does a guy without a criminal background or prior arrest say to himself, "I guess I'll go out and stick someone up." Every robber I've ever known, including myself, came up through layers of crime—car theft, reform school, burglary, prison, sometimes assault, and, finally, armed robbery.

Most armed robbers know that what they're doing isn't cool, and that it carries heavy penalties. But fear of punishment is absolutely no deterrent. They don't care. (I won't go into why, which is a book in itself.) Apart from this essential and fundamental recklessness, the other forces that motivate them are the desire for easy money, the power it gives them, and a need for the release of excitement.

What you need to keep in mind if you're confronted by an armed robber is that this turkey is dangerous. He's nervous and excited. *And he doesn't care.* Your best chance for living through the experience lies in remaining calm and closely obeying the following 7 rules:

1. Above all else, stay cool. Try not to panic or show signs of anger or confusion.

2. Don't argue or do or say things to cause delays. That includes cracking jokes, interrogating, bargaining, or even making pleasant conversation. Don't get in his way or attempt to slow him down. For your own safety, and that of the people around you, just give him what he wants and let him go. And don't give him a lecture. This is no time for social work.

3. **Listen carefully to his instructions. Do exactly what he tells you to do—no more, no less.** Repeat his instructions aloud to show that you understand and are cooperating.

4. **Don't make jerky or abrupt movements or look wildly around.** Say out loud what you're going to do before you do it. Tell him which pocket your wallet's in before carefully and slowly removing it between thumb and index finger (not gripped in your hand). Again, give him what he wants and get him away from you. The shorter time you and the others have to face that gun or other weapon, the better your chances of coming through unhurt.

5. **Never, never try heroics.** Don't grab for his gun. No gambler with more than three IQ points would take those kinds of odds—particularly when lives, not money, are at stake.

6. **Wait for him to leave.** When you've given him your money and valuables, wait calmly for him to leave. If he gives you further instructions, follow them. *The most dangerous moment in any robbery comes just after the robber has gotten what he came for.* Now he's wondering what to do next. He's suddenly got a whole new problem: How to keep you and the others from coming after him and thwarting his getaway. At this point almost anything, no matter how trivial, will set him off. So remain calm and do what he tells you.

One of the women employees of 7-Eleven handled this crucial moment very well. She had never been held up before, but she had been through a robbery prevention training program. She told the robber, "The last time I was held up, the man told me to lie down on the floor. Do you want me to lie down on the floor?" The young robber, who had been hesitating uncertainly, said, "Yes." She lay down on the floor and he was gone.

7. Don't follow the robber. When the robber leaves, don't follow him into the street or run after him. Call the police. Capturing criminals is *their* work.

When the police arrive, that's the time to be a hero. Because you kept a clear head and noted many important details, you're able to give them an accurate description—height, weight, coloring, hairstyle, outstanding features, dress, unusual marks or scars, speech patterns and accent, mannerisms, and anything that stands out. These are the things that will help most in identifying the robber and in making an early arrest to get him off the streets.

If you think this sounds like I'm making it easy for the crook, ask a cop. He'll tell you the same thing. You may not like giving a robber your money and valuables—but you better believe you'd like it one whole hell of a lot less if he blew a hole in you or started carving you up.

So, if you're confronted by robbery, *think*: Two scared people and a loaded handgun are a prescription for disaster. Don't give it time to happen.

Automatic Teller Machines
(See also Armed Robbery; Mugging; Personal Safety; Rape; Self-Defense.)

Just in case you've been away for a while (as parolees, say), automatic teller machines (ATMs for short) are those one-eyed, one-armed bandits banks are replacing people with. They look a little bit like computers, which is basically what they are, and you're going to be seeing a lot more of them. The banks are currently testing the machines in several thousand retail store locations around the country. If the tests prove successful, by 1990 you may have forgotten what the inside of a bank looks like.

Although there are ATMs based on different systems, their operation is usually the same. The bank issues you a plastic card with your ATM number magnetically imprinted. You put the card in the slot and punch in your transaction—withdrawal, deposit, payment, whatever—and the machine wishes you a good day. If it's in an especially good mood, it might even give you some money. Rules vary, but withdrawals are usually limited to amounts between $200 and $500 per transaction, with the machines programmed to accept only one withdrawal during a specific time period of 12 or more hours.

The machines, or rather their locations, can be dangerous. They're in public places, and people go to them to get money at all hours, sometimes late at night when the streets are dark and deserted. The danger is pointed up by one of Johnson's laws: "Where there's cash, there are going to be predators."

The bottom line is you have a fixed location in a high-crime area that's open for business 24 hours a day to which average, ordinary people regularly come to withdraw respectable amounts of cash—and every guy who owns a gun and has a drug habit within 50 miles in any direction knows it.

Of course, so do the police. But the police don't have the personnel to watch every single one of these things around-the-clock. The best we can hope for is an occasional spot check. During the times in between, the muggers just sit and wait for us to come draw out their money for them.

There's no arguing with the convenience of ATMs. But we as bank customers have a right to expect the bank to provide for our security when we use their automatic tellers. Here are some things for you to think about:

1. Insist that your bank back up police checks with private security patrol checks and that they advertise the fact that their automatic tellers are patrolled by special security.

2. Banks should also be encouraged to publicize to their customers (and to the muggers as well) the limits on the amount of money customers can withdraw from ATMs. This will reduce the number of times robbers go to the ATMs thinking they'll get more than they will and then become enraged with the victim because they think he or she is holding out.

3. All locations should be brightly lit at night. The lighting in and around the ATM I use occasionally is not good, and I've complained to the bank. If the ATM you use doesn't have bright lighting, write a letter to the president of your bank. When crooks have a choice between places offering the same opportunities for getting money, they'll choose the darker, less visible one every time. They love privacy.

4. Because we know the risk involved in using the ATMs, we have an obligation to be careful. As an extra precaution, I drive around my bank to make sure no one is lurking around the corner. I'm just as cautious as you should be. A lot of people I meet seem to think that all it takes for me to stop a thief is to say, "Hey, I'm Ray Johnson, the well-known excrook! Man, I did my time in prison!" The guy's then supposed to pocket his blackjack, shake my hand, and shuffle off to tell his friends. Bull! That turkey will knock my brains out as quick as he will yours. The truth is, I'm just as likely to be ripped off as anyone else. So I try to be careful.

5. If you usually drive to your ATM, take someone with you who can act as lookout. If you have a CB, keep it tuned to the emergency channel in your area that's monitored by your local police. Make sure your pal understands how to use it in case of trouble.

6. If you're alone and someone who looks suspicious walks or drives up or you see that you're being watched,

RAY JOHNSON'S TOTAL SECURITY

don't complete your transaction. Leave the money in the bank, get in your car, and go. You can always come back later. Your safety is worth more than anything you can get from the machine.

Automobile Trip Security
(See also Alarms, Auto; Armed Robbery; Business Traveler Survival Tips; Camping Trip Crime; Car Stripping; Car Theft; Freeway Crime Smarts.)

Getting there is half the fun—the other half is getting back in one piece. You can't have a safe trip without good planning. Plan ahead. Before taking your car on trips, provide it with necessary security.

1. **Service your car.** Replace any worn or weak parts that might contribute to a breakdown on a dark road at night.

2. **Install security devices.** These include alarms, ignition cut-off systems, and so on. Make sure you have hood and trunk lock security (described in the sections on "Alarms, Auto," "Car Stripping," and "Car Theft").

3. **Secure luggage.** If you're using luggage carriers or other car-top storage, secure them well with strong fasteners and good locks.

4. **Wire campers.** If you're taking a camper or pickup cap, make sure to wire it into your car alarm system.

5. **Buy a CB.** If you don't already own one, buy a reliable CB radio and a good antenna for your car.

6. **Check your tool kit.** Make sure you have the neces-

46

sary tire-changing tools, a dependable trouble light with fresh batteries (or one that will work off your car battery), and emergency flares in case you have to stop on the highway.

7. Plan your route carefully. If you belong to an auto club, get strip maps and road directions and ask for current information about construction, bridge closings, detours, etc. And make sure you have a good, reliable compass.

8. Avoid unfamiliar areas. Do not leave your route for unfamiliar terrain.

If your trip involves overnight stops at motels, here are some points to keep in mind:

1. Check out the motel. If you haven't had previous experience with the motel, check the place out before taking rooms.

- Apart from the usual amenities—color TV, bath/shower, free ice cubes—are there double locks, including thumb turn dead bolt, on the door?

- Can you park your car close to your room so it's in full view from your window?

- Is the motel situated in a reasonably clean and neat neighborhood? Does it appear well cared for and maintained? Are there unsavory characters in the bar or lobby?

2. Don't be afraid of moving on. If you don't like what you see or the place doesn't check out, drive on. There are a lot of motels, but there's only one of you and your family.

3. Keep valuables with you. When you do stop, don't leave valuables locked inside your car overnight—especially on the seats or dash where they'll attract attention. Take wallets, travelers' checks, cameras, expensive sunglasses, binoculars, toll change, keycases, luggage, and other items into the room with you. If someone steals your car, at least they won't get everything you own with it.

4. Keep your car alarm armed. Even if your parking place is directly in front of your door, keep all locks locked and the keys with you. There's nothing more embarrassing than having a thief slide into a loaded car, armed to the teeth with security, and drive off because you left the key in the ignition or failed to set the alarm.

5. Don't leave your car keys in the room. Keep your car and room keys with you at all times. This is a good idea whether you're in the restaurant, lounge, or at the pool. Remember, keep your car keys, travelers' checks, credit cards, cash, etc., on your person. If swimming, leave them where you can keep an eye on them or with someone you know.

6. Do not take furs and jewels. Discourage anyone in your party from bringing along expensive jewelry, furs, electronics equipment, cameras, or other items that might attract street thieves and other criminal types.

7. Keep your hotel door locked. This is true whenever you're in the room, as well as out of it. If callers appear unannounced with unexpected room service or messages, tell them to leave it by the door. Be careful of strangers. Push-ins in hotels and motels are on the increase.

B

Bunco
(See also Con Games; Elderly, Common Crimes Against the.)

Bunco is a kind of con game that involves sleight of hand. Bunco literally means flimflam, a way to fleece people without overt threat of violence, usually by trickery. It includes everything from three card monte to the pigeon drop.

BUNCO-TYPE GAMES

Three Card Monte

Three card monte is a variation of the old shell game, in which three walnut shells and a dried pea or bean are used. The operator bets you that you can't pick out the shell that the pea is hidden under. In the variation using three playing cards, three cards of the same value—kings, jacks, aces, or whatever—are shuffled around on the playing surface, usually a milk crate with a handkerchief over it or the bottom of a cardboard carton. You're challenged to pick the king of hearts in a $1, $2, or other cash wager with the operator. He usually lets you win a couple before he really takes you for your bundle.

Pigeon Drop

The pigeon drop is not a single routine—it's a whole class of scams. They're usually run in urban areas by well-dressed, average-looking characters who are highly skilled at their trade. They're so convincing that, in spite of crime-prevention campaigns to educate the public, thousands of people, both young and old, lose their savings to them every year.

There are minor variations, but most pigeon drops follow this general scenario:

You're sitting at a bus stop waiting for your bus. Someone who looks like your aunt or uncle in Peoria sits down beside you and says, "Pardon me, what time does the next bus come by?" Before you can answer, the person's gaze drops, skillfully directing your attention to an envelope at your feet. "Is that your envelope?" the person asks. Being an honest citizen, and not wanting to get involved, you say it isn't. The person then picks up the envelope and opens it, letting you see that inside there is $5,000—five new crisp thousand dollar bills.

The person slyly presses the money on you, saying that after all, it was *you* who found it. You protest weakly, still in shock at the sight of so much money, feeling confused and a little guilty.

About this time, the person's confederate shows up, posing as another solid citizen waiting for the bus. He's been lurking in the background, following the play to make sure everything goes right, ready to intervene and abort the scam if things go wrong. In his role as innocent bystander and commentator, he soon learns of your apparent good luck and resulting quandary. He makes a helpful suggestion. Since you and the cofinder are strangers, obviously *both* of good character and *both* involved in the finding of the money, why doesn't one of you put up good faith money—a half share, or $2,500—then bank the $5,000 for safekeeping while you wait to see if anyone claims it? In the unlikely event that this should happen—since there's no way to identify cash in circulation—whichever one of you puts up the good faith money would then get it back. You begin to calculate the interest you'll get on $5,000 in your super saver account.

Your cofinder is suitably depressed, reluctantly admitting that he doesn't have $2,500 in the bank. It's up to you. You go to the bank, draw out the $2,500, give the person the money, and receive the envelope in return. You open the envelope and your hand comes out with five brand-new strips of yesterday's newspaper. You look around for your friend, wondering half aloud: "Who was that stranger?" but your $2,500 is already headed for another bus stop, in another part of town.

Punch Board Schemes

Phony punch board schemes are another type of bunco. These are board games in which you punch out tiny rolls of paper or flat dots that are alleged either to be blank or to have various amounts of money printed on them. You pay the operator anything from a quarter to several dollars or more for each chance you take. The rules of the game prevent you from buying more than 75% of the spots on the board. Of course, you never get the one with the money on it. It isn't there. It's in his pocket. If you challenge him, he quickly palms it, punches one of the spaces out of the board, makes the switch, and shows you the piece with $500 printed on it from his pocket, just to prove—with a contemptuous sneer— that it's in there.

HOW TO AVOID BUNCOS

What can you do to avoid being taken in by bunco? Another Johnson's law: There's no such thing as a free ride. Believe it, and you'll never be a mark for bunco.

Here is some advice for specific situations:

1. Found money. If you are ever involved with other people in finding money, insist that they accompany you to the nearest police station. Turn the money over to the police and have them give each of you a receipt for the full amount, with the names and addresses of each of the finders. When can you collect the money? According to most city codes, finders are entitled to money not claimed by people who can prove ownership beyond reasonable doubt within time limits specified by law. So find out what the time limit is.

2. Wagering; games of chance. Don't get involved in wagering or games of chance with people you don't know. When I'm in New York, I often see tourists crowding around some turkey who's dealing three card monte. In some towns, you find these and other scams operating openly. Don't pay any attention to them; just walk on by.

3. Making change. Don't agree to make change for strangers in a crowd or on the street. One currently popular scam in the cities is for bunco artists to approach their victims with a $20 or $50 bill, saying they've got a cab waiting to get paid or some other urgent need for change. If you reach for your wallet, you're lost. Through a few artful and polite but hurried maneuvers designed to confuse, they wind up taking their bill back and disappearing. Later you find you're short of cash.

The success of these and other bunco scams rely heavily on their appeal to our greed, egos, and feelings of guilt.

Not long ago, a friend of mine, a successful business guy who happens to be black, was ripped off in a variation of the pigeon drop by a black couple who drew him in with the "less fortunate black brothers" routine. Some of these hustlers have a highly developed sense of smell for "good samaritans." Tip number 3 above illustrates how a willingness to help can sometimes lose you your money.

I've barely scratched the surface in pointing up the most common forms of bunco; there are many more, with new ones hitting the street everyday. I don't know of any absolutely certain, 100% sure way to avoid bunco in all its variations, but you can cut your losses dramatically if you ignore attempts by strangers to draw you into situations promising quick money. Another good rule is to avoid any situation on the street or in public that involves your reaching for your money. Leave your money in your purse or wallet.

Burglary
(See also Alarms, Home and Business; Apartment Security; Doors; Home Security; Locks; Windows.)

I did my first burglary when I was still young enough to consider teenyboppers the big guys. I'd just escaped from a juvenile home, and since all I'd got away in was my warts and a kind of reversible nightshirt they kept me in 24 hours a day to discourage escape attempts, I was looking for clothes. It didn't take me long to find an unlocked house with some kid's clothes in it.

That was a long time ago—another life—but you'd be surprised at how many of your neighbors still don't lock their doors, even when they go out for long periods of time. If I were to walk down that same California street today, I'll give you odds I wouldn't have to ring more than a few door bells before I found a house with nobody home and the doors unlocked.

Like much of the crime I talk about, most burglary can be prevented. Sure, there are the exceptions where determined burglars go to a lot of trouble to bypass elaborate security to make a score. But these are the guys who know what they're after, and it's usually something big—collections of valuable jewelry, coins, art, large sums of cash. These are the professionals, like my friend, the Bel Air Phantom, I told you about in the section on home alarms. Their kinds of jobs represent a tiny percentage of burglaries. Well over half of the very high residential burglary losses we suffer every year are committed on the property of your neighbors and mine, people who have been careless and failed to take simple, inexpensive precautions.

10 WAYS TO REDUCE YOUR CHANCE OF BEING BURGLED

The point is, you *can* reduce your chances of being burgled by at least 50% simply by seeing to it that your house has good, bright lighting indoors and out and looks lived in, well cared for, and protected. Here are some specific precautions you can take:

1. Take a long, hard look at your doors, windows, and their locks. One of the things I advise people to do in crime prevention meetings around the country is to try a little do-it-yourself housebreaking. Lock yourself out and see how long it takes you to get in without your key (but keep your key in your purse or pocket in case you find that your security is too good). If you live in a highrise, this may not be so easy. But if you live at ground level, you'll probably be surprised at how simple it is to get in. Most of our homes and lower floor apartments can be broken into without even smashing a window. That's too bad, because that's just what your amateur burglar—the most common kind—counts on to get into your house or apartment.

If the locks on your doors and windows don't work, replace them. If they're weak or of poor quality, there are some things you can do. (You'll find suggestions in my entries for "Alarms," "Doors," "Locks" and "Windows.") Remember: By simply keeping doors locked, we can reduce the number of burglaries by one fifth and cut our losses by over half a billion dollars.

2. Check your outdoor lighting. If you live on the ground level or second floor of an apartment building, insist that your super or apartment manager keep the area outside

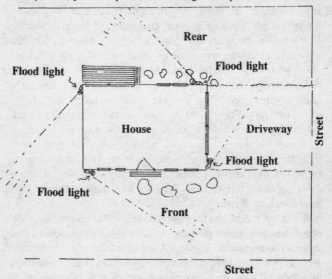

Figure 5. OUTDOOR LIGHTING.

your apartment well lighted. If you live in a house, have an electrician do it, or do it yourself. There may have been a time when we could get by with a 40-watt bulb in the porch light. But times have changed, and we need to change with them.

For a single-family residence, the ideal lighting is floods at each corner of the house. For tips on how to do it yourself, see Figure 5. Floodlight fixtures should be metal or porcelain. Fit them with 100-watt outdoor-type reflector floods. If you've already installed a burglar alarm, make sure you wire in your outdoor lighting so it automatically goes on when the alarm is triggered.

3. Install a burglar alarm. Preferably, install a good quality, perimeter alarm. A good alarm, properly installed and maintained, is your best 24-hour protection against burglary. It will ward off all but a very small percentage of burglars. A lot of people try to minimize the value of burglar alarms: Don't buy it. Burglary is a crime of stealth. No one, whether kid or professional, is going to stand around with a bell or siren going off around him. And contrary to the popular misconception, you don't have to spend a lot of bread for a good alarm system. The average house or apartment can be protected for under $100 if you do the installation work yourself or if you have a friend or relative do it for you.

Home burglar alarm manufacturers estimate that fewer than 1% of the 90 million dwellings in the U.S. are currently protected by burglar alarms. Last year, some 4% of those dwellings were ripped off—I'll give odds most of them didn't have working alarms.

4. Don't Kid the Burglar. One of the gimmicks mail-order hustlers keep pushing is the old FBI window-label scam. The copy tells you that for five bucks you can receive 10 ''official'' labels to put on basement and downstairs windows, doors, etc., warning burglars that your house is protected by the FBI, Treasury Department or some super-duper burglar alarm system. The idea is to impress your taker with the futility of trying to break into a house with all the impregnable defenses you don't have.

The problem is we all get the same junk mail—including the burglars. And nobody with any sense takes these warnings seriously. In fact, you're more likely to make your property more interesting to intruders by putting up signs. As long as you're just one in 10,000, you've got anonymity going for you. Start sticking signs all over the place and you suddenly seem concerned—and important.

5. Don't hide behind shrubs. Don't kid yourself into thinking you can hide your family and yourself behind a bunch of shrubs, trees, and hedges. Burglars and robbers are private people, too. Keep the greenery trimmed back so that passersby, police, and neighbors have a good view of your house.

6. Secure your bedroom. Make your bedroom the nighttime security center of your house or apartment. Harden doors and other possible entries. Install a dead-bolt lock on the inside of your bedroom door. Pick up a wireless, remote lighting control center from an electronics outlet. The main control console, with one or more rows of buttons, is small enough to fit on your bedside table. Simply plug it into a wall outlet in your bedroom, plug lamps and appliances into the remote modules, and replace conventional wall switches for outdoor lighting with two-way remote selective units (see Figure 6). Then, by pushing the buttons on the control unit at your bedside, you can turn on any light or appliance inside or outside your house from the security of your safe room.

There are several types of remote lighting centers currently on the market, and many models have timers built in that can be set to switch various circuits on and off at seemingly random times.

I have a story I tell about these machines. When I was married both my wife and I worked, so we had a woman come in to do the housecleaning one day a week. We were talking about my doing a commercial

57

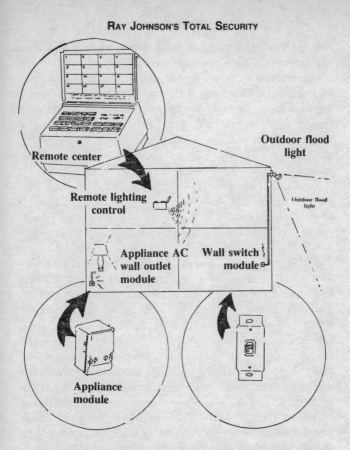

Figure 6. REMOTE LIGHTING CONTROL CENTER.

for one of the companies who make the machines, and they loaned me one to test at home. I hooked it up and forgot about it. The next day I left the house early. When I got home in the afternoon, the front door was open, the vacuum cleaner was standing in the middle of the room with its motor running, and the TV, the stereo, and a couple of lamps all were busily blinking on and off. I had forgotten to turn off the timer on the lighting center. Later, the housecleaning woman explained that she had plugged in the vacuum and started to work when the room just sort of "exploded" around her.

7. Put lamps on timers. Use simple, plug-in timer switches to turn on lamps in different rooms while you're at the movies or out to dinner. Hook one up to turn your TV on and off so eavesdroppers will think you're at home.

8. Secure fire escapes and skylights. If you live in an apartment with a fire escape or skylight, be sure to secure entry from intruders with some form of alarm. There are several door and window electronic alarms suitable for this purpose (see "Alarms, Home and Business").

I don't recommend the lockable steel gates some people install, because they're dangerous in case of fire. But if you just happen to live next door to some of the dudes I have, it's probably more dangerous not to have them. If you do install gates, be sure you, as well as everyone who lives there, knows the lock well enough to open it quickly. Mark the key plainly and leave it someplace where family or guests can easily find it.

9. Close those windows. When doing work around the house, don't go off and leave doors or windows open or ladders up and tools out, even if you're just running down the street to the hardware store.

10. Exchange house keys. Give a neighbor your key and get his. Show him how your alarm works, too. Get him to show you his. Set up a neighborhood watch plan for your mutual benefit. When you go on vacation, ask the neighbor to check your house from time to time; when he goes away, do the same for him.

Report any strange vans or unscheduled visitors to the police. Collect mail and newspapers and hold them. Depending on the season, mow grass, rake leaves, or clear snow so that your houses look lived in. If no neighbor is available, hire a reputable house-tending service.

SAVE A KID

Of the 2.2 million residential burglaries reported in 1982, the majority were daytime burglaries into the homes of middle-class working people. Over 40% of these burglaries were done by kids under 18, kids from the neighborhood or kids who live within five miles of us. And they're not all poor kids. Believe me, I know. I wasn't a poor kid either—but I still spent 25 years in the joint. Talk to your kid about it, even if you think burglary's not his style. Maybe he, and other kids like him, can talk some neighbor kid out of getting into trouble.

By maintaining good home and apartment security, you can reduce the opportunity for these kids to turn into burglars. Since most of them can't pick locks, bypass alarms, or crack safes, you'll probably get rid of most potential burglars under 18 and give a kid a chance to grow up into a law-abiding citizen.

WHEN BURGLARS STRIKE

The odds that you'll be at home during a burglary are pretty slim. Most residential burglars strike when you're out. But just in case, here are a few tips:

1. Turn on the lights and call the law. If your burglar alarm goes off in the middle of the night, turn on all the lights and call the law. Don't go poking around in the bushes on your own. Let the police do it.

2. Shout! When you've neglected to install the alarm you bought last fall or to fix that window the kids broke and you hear someone big and ugly rattling around in your family room, don't start whispering and tiptoeing around. Sit up in bed and shout, "Honey, there's a burglar downstairs! You get the shotgun and I'll call the police!" Chances are the sound of your voice will send him out the window. Most burglars are just as anxious as you to avoid an introduction. If you give him time to get out of the house, he'll be on his way.

3. The phones are for you. Don't leave extension phones out where a burglar will have easy access to them. If your telephones don't already have universal instant jacks, 4-prong plugs, or module receptacles so you can unplug them and lock them in a secure place at night, have them installed (see "Telephone Security"). Extension telephones left out in the living room, family room, or other low-security area make you extremely vulnerable. The first thing most burglars will do if they get inside your house or apartment is lift the phone off the hook to prevent you from using your telephone to call for help. The last thing you need is to pick up your phone to call the police and find the burglar talking to his wife on your telephone!

4. Keep on driving. If you pull up in front of your digs after a night on the town to find a van or other strange vehicle in your driveway, just keep right on driving until you find a phone. Then call the police.

5. Avoid contact. Violence during contact is on the increase. Be alert. If you arrive home from the bridge club to find locks or windows broken or open, lights on in the wrong rooms, or lights off that should be on, be suspicious. Don't go in. Go to your neighbor's house and make a phone call. Get the police on the case.

Business Travel Survival Tips
(See also Airport Crime; Automobile Trip Security; Credit Cards; Freeway Crime Smarts; Hostage Situations.)

My work as a crime prevention consultant for The South-land Corporation keeps me on the road a lot. But for 25 years—or roughly from the mid-1940s to the late 1960s—I didn't travel much. The only business-related security problem I had during that time was how to keep the homicidal dude in the next cell away from me.

Those years in prison wised me up to the robbers who prey on business men. I heard stories from crooks who had worked hotels, burglarized rooms, and ripped off guests. Some of the more violent types rolled (mugged) business men, often working in cahoots with prostitutes who set them up. Others worked airports for luggage, cameras, and so on.

11 TIPS FOR SURVIVAL

Now when I travel, I have a healthy respect for the risks in business travel. I'll share a few of my rules with you:

1. Avoid mogul madness. If you can, avoid the dread executive affliction—mogul madness. Don't stride around the corridors of large convention centers and strange hotels as if you were about to fire everyone and replace them with computers. That's fine for corporate headquarters, but keep in mind that a lot of dudes running around loose don't understand how important you are or don't have the respect for your outstanding qualities that your fellow employees have. In other words, don't draw unnecessary attention to yourself.

2. Avoid arguments or confrontations. Don't hassle with waiters, bartenders, exhibit carpenters, electricians, and so on. If you don't get the service you or your company expects, talk to the manager. Don't hop on these people yourself. You don't know where they're coming from.

3. Don't be careless. Don't become careless or over-confident in handling your briefcase, attaché case, clothing bag, and other luggage. Everyone knows the feeling of having too much stuff to carry around, being late for a flight, and having to make an emergency pit stop at the rest room. It's tempting to pile everything up outside the door while you retire inside. When you emerge, the pain is gone and so is your stuff. So be sure to keep your things in sight at all times.

4. Avoid the brinkmanship syndrome. This is a common disorder of the executive mind. Haste makes you careless, confused, and preoccupied—and accident prone. In the mad rush, you fail to notice things that are lost or stolen—luggage, wallet, purse, credit card, etc. Give yourself plenty of time to get to the airport. A lot of problems in travel happen when we're trying to make up lost time. Don't putter around your office dumping ashtrays or reading the *World Almanac* until 10 minutes before flight time, then assault your cab driver, mug innocent bystanders, and leave a violent wake through the terminal in your dash to get to the plane before they close the door. Watch the time and start early.

5. Plan—or rather don't plan—celebrations. Don't celebrate the night before you leave on your business trip. Avoid excessive partying while you're on the trip, especially when you're traveling alone. It's tempting to get a little too relaxed after a hard day at the regional office and start looking for some action. Apart from all the obvious risks involved in being in a strange place, there's the problem of getting through the next day. After brinkmanship, tired blood probably contributes as much to security risk as any other affliction of the corporate body. Like the amount of time you allow yourself to get to the plane, the rest you get and the edge you give yourself out there against the competition is something over which you have full control.

6. Even staying in your room can be hazardous. Regardless of how good a reputation your hosts may have, don't leave hotel room doors unlocked while you're in the room. Whether you're working on that proposal, sleeping, or watching TV, keep the double lock on. There's usually a dead bolt in addition to the knob lock. Keep them both locked. You can also keep the chain on, but don't rely on it to keep anyone out if you unlock the door.

7. Take along backup door security. I solve the problem with one of those little 49¢ rubber doorstops—the wedge-shaped ones. When I have to stay in a room where the locks aren't so terrific, or even when they are, I jam the little rubber wedge tightly under the door for added security.

8. Use other gadgets. You can use the following as backups: a type of battery-powered personal alarm that converts to a door alarm, sold at most electronics stores for under $20; a portable smoke alarm you can carry in your flight bag and put up in your room to warn you of fire.

9. Don't be the host or hostess. If you do find yourself partying a bit after a hard day, don't move the party to your room, especially if you're alone and have just met the people for the first time. If you're a little unlucky, the cute blonde and her not-too-bright husband will turn out to be professional bunco artists who wind up with your cash and credit cards. If you're very unlucky, you could be in real trouble.

10. Watch out for hookers. Be especially wary of members of the opposite sex who travel in pairs and offer exquisite earthly delights. Once they're in your room, one of them keeps you busy while the other relieves you of your cash and credit cards. What you get usually isn't worth what they get.

11. Don't confide in bar friends. Your business title, company name, financial status, and other vital statistics are private, not for those you meet at the bar. They may get the foolish idea your boss would pay something to get you back.

OVERSEAS BUSINESS TRAVEL

Given the locations of major foreign business centers and the much stricter governmental control in many of these places, international travel is probably safer than traveling around the U.S. Nevertheless, most of the same precautions apply to foreign business travel as to domestic, and a few important additional ones are in order. For example, in the Mideast, Africa, Latin America, and certain parts of Europe, the chances of your becoming a victim of terrorists or of being held hostage are much higher than they are in the U.S. The chances of your being mugged or robbed in most foreign cities are less than they would be in your own city.

Tips for Reducing Overseas Risks

To reduce risks while you travel overseas, do the following:

1. Do some homework. Before leaving on a foreign business trip, learn as much as you can about the places you'll be visiting. Study street and road maps of surrounding areas; obtain as much current data as you can on the people and their culture, and pay particular attention to political or religious controversy. Get to know all the players by name. Identify groups and individuals that could spell trouble for you.

2. Get help with reservations. Let someone familiar with the places you'll be visiting reserve your accommodations, driver, and translator. Stay only at hotels they recommend, and travel only on those routes suggested.

3. Maintain a low profile. This is especially true if you're in one of the hot spots. Stay off the streets. Eat and entertain at your hotel. If you must travel between your hotel and an office or plant, follow a different time schedule and route each day. Stay alert. Keep an eye out for cars that appear to be following you. If you see anything that makes you suspicious, tell your driver to step on it. It's a lot better to get pulled over for speeding than to spend the next 10 months of your life in someone's wine cellar, even if it is stocked with Chateau Pétrus.

4. Don't play detective. Don't go off on free-lance fact-finding trips.

5. Don't debate. Don't engage in philosophical, political, or religious debates with anyone. Be courteous, but keep your trip strictly business.

C

Camping Trip Crime
(See also Automobile Trip Security; Freeway Crime Smarts; Personal Safety.)

Outdoor recreation and camping have become popular. After personal and business trips, they are the next largest reason for travel.

Because so many people go camping these days, including a lot of kids, and because of the nature of camping, which isn't conducive to strict law enforcement, the activity attracts a lot of vandalism and petty larceny. There's also a lot of not so petty larceny, and, unfortunately, growing incidences of violence as well.

TIPS FOR NONCRIME CAMPING

Here are some suggestions for avoiding crime or preventing it from spoiling the next outdoor outing you and your family take:

1. When planning your recreational trip, include plans for the security of the property you'll be taking along—tents, boats, trailers, etc. With a little foresight, and by leaving a

67

few inessential things at home, you can reduce the chances of your being robbed. For example:

- Do you really need to take your best tackle for the kind or amount of fishing you'll be doing?

- Should the kids take their best, most expensive stereo? Or will an older, beat up, portable radio do?

- Are you really going to get enough use out of the boat and outboard to haul it all that way—and take the risks involved? Or would you be better off renting one for fishing?

- How about cameras and binoculars? Will they be secure in the car? Can you take old or inexpensive equipment?

- Do you really need to take your best gold watch and jewelry along on a camping trip?

2.   When purchasing equipment for your recreational needs, plug a new factor into your thinking. I call it the possible loss factor (PLF). You arrive at it by calculating your ability to keep the equipment safe when it's not in use (i.e., locked, stored indoors, nailed down) and safe when you're on a trip (i.e., stored on top of your car, like skis, or inside the trunk, like boots).

Here are the types of questions you should ask yourself:

- Is the propane stove any less likely to be snatched than the kerosene one? Can one be stored more securely than the other? If not, which would mean the greater financial loss?

- Is the family-size tent from the expensive sporting goods store any better than the one from the

local discount chain? Is one more secure than the
other? Does the difference justify the price and
the risk?

Admittedly, this probably represents a considerable change
in your way of thinking and planning for recreation. But it
reflects the very real possibility that, when you and your
family embark on a recreational trip today, your chances of
having something stolen or vandalized are higher than ever
before. Once you've accepted this and taken it into consideration in your plans, you'll reduce the probability of it happening to you, as well as the amount of loss you might suffer if it
did occur. You and your family will breathe more easily, feel
more relaxed, and enjoy the outing more.

3. Special precautions should be taken when you're participating in potentially dangerous outdoor activities like hunting.
When not in use, guns should be kept locked securely inside
the car or camper, preferably in a trunk with a lock. Each gun
should be broken down and locked inside its own case,
without ammunition. Ammunition should be locked in a secure place separate from the guns. When you're hunting,
carry an unloaded gun, with the breach or bolt *open*. If
you've practiced loading your piece, you'll rarely lose a good
shot in the time it takes you to load. Even if you do lose one
now and then, it's better than losing your head while ducking
under the barbed-wire fencing.

4. If your budget (and group consensus) permits, rent a
vacation motor home to drive to the mountains or shore.
These offer maximum recreational security yet will go most
places accessible to campers. If your party insists on roughing
it in a tent, you can attach it to the motor home. If you're
backpacking in the wilds to get away from the crowds, a
motor home with adequate locks and security makes a reasonably safe storage place to leave equipment you can't take with
you.

5. Be sure all recreational equipment; pop-up campers; boats; motors; generators; stoves; kerosene, gas, or LP lanterns; tents; sleeping bags, etc., either have locks or can be placed in something that can be locked when they're not being used.

6. Take several electronic personal alarms (see page 173) along on your next camping trip. These make excellent, jerry-built outdoor alarms for protecting camping equipment. They're also good attention getters if anyone gets lost in the woods or runs into trouble with other campers who have more on their minds than tenting out.

7. Don't leave your locked motor home or other vehicle parked in a camp ground, at the side of the road, or off in the woods for more than a few daylight hours. Even then, it's best to check on it frequently. Keep in mind that regardless of locks a determined thief can break into it in a matter of minutes. Park in an area designated for recreational vehicles, and offer to pay someone to keep an eye on it for you. If there is no such area, you may be able to find a service station that will let you park in its lot. In some resort areas, there are motels or tourist courts where campers can pay a modest fee to park their recreational vehicles.

Car Stripping
(See also Airport Crime; Alarms, Auto; Car Theft; Larceny/Theft.)

Not long ago my neighbor left his car in the driveway overnight. When he got into it the next day, the dashboard was gone. Have you ever tried to take the dashboard out of your car? I have. It isn't easy under the best conditions. The jerk who took my neighbor's dashboard must have worked all night, hanging upside down over the front seats like an orangutan with his head up under the dash. But that's the kind of thing car strippers do. They're mostly neighborhood kids. Professionals simply steal the whole car and take it to a chop shop.

If you live in a high-crime area, chances are you've already lost something—a set of windshield wiper blades, wheel covers and wheels, a battery, a spare tire—to "Midnight Auto Parts." Whatever you've lost, chances are you could have discouraged or prevented the theft if you'd taken certain precautions.

WHAT'S YOUR CHANCE?

Take the following quiz to determine your chances of being victimized by a car stripper.

	Yes	No
Is your car less than two years old?	___	___
Is it an older car in exceptional shape?	___	___
Do you usually park it on the street or in an unattended lot?	___	___
Are your parking places generally on side streets or where street lighting is poor?	___	___
When you take your car in for mechanical work or service, is it left out overnight or on weekends?	___	___
Is your car a rare model, popular classic, or collector's item?	___	___
Does it have custom wheel covers, wheels, and tires?	___	___

	Yes	No
Have you put off installing a car alarm or neglected to install pin switches for the hood and deck lid?	___	___
Does it have a CB radio and antenna, digital stereo, power amp, or other goodies?	___	___
Total responses	___	___

If you scored higher than zero in the yes column, you're a candidate for stripping. If you scored higher than two or three, you're probably in immediate danger.

HOW TO STOP CAR STRIPPERS

Just keeping your car's trunk locked isn't going to stop the stripper (probably a kid) who's after your new spare. All he has to do, on most new cars, is give the trunk lock a good whack with hammer and punch—and he's inside. It's a 10-second operation. But most car stripping can be avoided. All it takes is a little thought and a willingness to make the additional investment in time and money.

Tips to Prevent Stripping

Here are a few tips on how to harden the target for your neighborhood car strippers.

1. Protect your trunk. For under $40 you can have a locksmith harden your trunk lock with a heavy-duty anti-theft plate and lock assembly. It's well worth it, even if all you're ever going to carry in the trunk is your spare tire, jack, and tire-changing tools. If you've ever found yourself late at night on an interstate with a flat tire and discovered there was no spare tire in your trunk, you know what I mean.

2. Protect those extras. Make sure your stereo and CB are protected by a good quality car alarm and are well secured.

If you install the equipment yourself, use heavy-duty special fasteners requiring Torx or other odd tools that thieves are less likely to be carrying. You may have to get your auto parts store or hardware store manager to order them for you, but most stores have catalogs listing special fasteners, as well as tools required for installation. If you're having the equipment installed, ask the installer to use similar fasteners or, if nothing else, to peen (flatten the ends) the fasteners to make removal difficult.

3. Install hood locks. If you don't already have them, install hood locks and make sure your car burglar alarm system includes pin switches protecting both hood and deck lid. Pin switches are among the least expensive components of any alarm system, and those you install in the hood protect some of the most frequently stripped parts such as the battery.

4. Don't let them move your car. To prevent busy little elves from jacking up your car and stealing tires and wheels, your alarm system should include a mercury switch to trigger the alarm if the car is jacked or moved. Some manufacturers describe them as motion sensors or motion detectors and alarms. These gadgets will also secure your car against car thieves using tow trucks.

5. Safeguard your wheels and tires. If you care enough about your new tires and wheels to invest another hundred bucks or so, you can pick up a set of lockable wheel covers or locking stud nuts at your accessory store. While they're not foolproof, they will deter most amateur car strippers.

6. Park for protection. If you don't already have garage space for your car, run an ad in your local paper for lockable garage space within convenient distance of your house or apartment. If none is available, when you do park take a few extra minutes to drive around until you find a parking space in a well-lit, highly visible, and well-trafficked place where thieves will be less likely to strip your car and where, if the car *is* being tampered with, your alarm is most likely to attract attention.

Identifying Car Parts

Law enforcement's major problem in dealing with theft is identification. Parts that are stripped out of cars are seldom identified or accounted for. For years, legislators have been looking for answers. One that's not popular with car makers is the idea of installing consecutive digit stamping equipment on their tools, which will stamp identification numbers on parts as they're manufactured. The manufacturers argue that the costs of such a system would increase the price of cars considerably. Those in favor of the idea point out that car makers have already lowered their costs by using cheaper designs, cheaper materials, and automating plants and that only minor modifications of existing tooling would be required. They suggest that car manufacturers aren't eager to help the fight against car strippers since stripping actually results in more sales for them.

If you really want to stop car strippers, I've got an idea worth passing along to your state legislator, representative, or senator. Let's pass a law in every state requiring car makers to number consecutively all parts with permanent, invisible, magnetic ink that can be recognized by equipment similar to that used in sales and checkout counters. Most crooks (that is, those without access to the equipment) wouldn't be able to see the numbers, so how could they remove them?

With this law in effect, every car or part recovered would be identifiable in minutes, no matter where in the country it was stolen. When the cops nabbed some kid with a garage full of stuff, you'd get your's back. More importantly, auto theft and chop shop operation convictions would zoom, and every thief in the car parts business would start thinking of finding steady work in some junkyard.

Car Theft
(See also Airport Crime; Alarms, Auto; Camping Trip Crime; Car Stripping; Vacation Travel Tips.)

Half of the over one million cars and other motor vehicles stolen in 1982 were taken by kids, and 57% of the people arrested for auto theft were under 21. Unfortunately, kids aren't fussy thieves. Their objective is usually nothing more than a joyride (which all too often ends in tragedy), and they aren't particular about what kind of car they take or where they take it from—be it your driveway, the street, the parking lot, anywhere.

YOU CAN PREVENT CAR THEFT

Because many of us have the habit of leaving our keys in the car, we actually make it easy for young car thieves. Police and insurance records show that one out of every four cars stolen had the keys in the ignition. That's a foolish, costly habit. If we could break ourselves of it we'd prevent over 250,000 car thefts a year, keep a whole bunch of kids from ever having to see the inside of a juvenile court, and maybe even convince our legislators to roll back auto insurance rates. That's a small price to pay for such a high return. So make it a habit to take your keys with you and lock your car behind you, even if it's parked in your own driveway or garage.

But there's an even better reason to make an effort to reduce the number of young car thieves. Many of these kids are launched into lives of crime by our carelessness. I know because stealing cars was the way I got my start in the life of crime, which resulted in a million dollar "hotel bill" for California taxpayers and that took an expensive 25 years out of my life.

Just remember: Whether the theft of your car launches a kid on his way to prison, gives a robber a lift to the bank, or contributes to the Mafia family fund, your failure to do what you can to prevent it costs you and your neighbors a whole lot more than just the replacement cost of your car. And the cost may increase because many states are looking for ways to crack down on negligent drivers. For example, New York State law now makes you liable for damage incurred by illegal use of your car if it is stolen as the result of your leaving the key in the ignition while the car is unattended.

STEALING CARS IS NOT JUST KID STUFF

Although kids represent the largest number of car thieves, it is the adult car thieves who pose the greatest threat to car owners. Most of the cars recovered are those stolen by kids; those stolen by adult car thieves are rarely seen again in one piece.

There are four kinds of adult car thieves:

1. The guy who's in a big hurry to meet new people and needs your car to get away from the old ones

2. The crook who needs wheels that can't be traced back to him to use for transportation to and from the place he's planning to rob

3. The members of car theft rings

4. The operators of chop shops, where your car is completely disassembled and sold as parts

An increasing number of adult car thefts are the work of professionals who fall into the last two categories. The cars that fall into the hands of these pros are seldom recovered. In the case of the chop shops, your car dissolves into a bunch of anonymous auto parts within hours of its theft. Your best defense is to keep these pros away from your car in the first place.

WHOSE CAR IS IN DANGER OF BEING STOLEN?

Car theft is primarily a big city problem. It does happen out in the boonies, but it's a whole lot easier to deal with it in places where everyone knows who's driving what.

In general, the more expensive your car is, the more likely you are to have it stolen. If you're among the majority of car owners who live in metropolitan areas with populations over a million, the chances of having your average four-year-old car stolen are about 1 in 100. The chances increase if your car is rare or special in some way; for example, if it's a sports or performance car, a vintage model, or an antique car. And if your car is new, especially if it is a luxury or sporty model, the chances of your having it stolen are about 1 in 40.

Changing the Odds in Your Favor

It's not hard to swing the odds in your favor once you've made up your mind to do it. The best insurance against car theft is still your own determination not to let it happen to you. With the right mental attitude and reasonable security precautions, you can lower the probability of having your car stolen to almost zero. Here are some specific suggestions:

1. Take the edge. Whenever you leave your car, regardless of how long you'll be gone, lock it and pocket the key.

2. Disable your car. As an extra precaution, temporarily disable your car by removing the distributor rotor or some other small essential part. (Your mechanic can show you how.)

3. Install protective devices. If you don't already have them, install hood locks, a good passive alarm system with ignition cutoff, and other antitheft security devices. Protect your security investment with frequent tests and regular maintenance.

4. Lock it away. Keep your car in your locked garage when it's not in use. If you have no choice but to leave it on the street, park it in a high visibility area, one that is brightly lit at night.

5. Avoid leaving it alone. Don't take your car to the airport or other places where it it is likely to be left unattended for long periods of time. Have someone drive you, or take the bus.

Computer Dating Services
(See also Personal Safety; Singles Bars.)

Computer and video dating services like to tout themselves as high-tech matchmakers. The truth is they very often don't use computers and sometimes offer very little service. And although some of them claim a high success rate in bringing people together in ''meaningful relationships,'' others make a good buck churning you through their list of well-heeled one shots and a few are simply fronts for hookers of both sexes.

Dating services are usually small businesses, with a dozen or so employees. Many have no assets other than their typewriters, equipped with computer-style type bars or elements, and some office furniture—and sometimes even this is rented. Apart from the computer time leased by a few of the larger companies and franchised chains, their major capital outlay is for the advertising that gets you and their other clients in the door. I don't mean to imply that they're fly-by-nights; *I mean to say it right out*.

Most of these services insist that you provide some very personal information, such as financial situation, family history, and several recent photographs of yourself. Some of them interview you on video tape to be shown later to people interested in your type. In turn, you are allowed to select from among a carefully screened group of prospects taped in similar interviews.

Not that I'm against love. Man, I'm in love with love! If you don't believe me, ask any of my 8 ex-wives. But if you'd done time with some of the dudes I have, I guarantee you'd think twice about putting yourself in the hands of strangers.

One of the inmates in Folsom prison was a real pussycat. He had the most angelic smile you've ever seen. And when he was smiling, he was no more dangerous than a King Cobra about to strike. But twinkly eyes and all, he was doing consecutive life sentences for murdering a string of his girlfriends from Miami to San Luis Obispo. Some dangerous things come in pretty packages.

LOOKING OUT FOR MR./MRS. GOODBAR

If indeed it's true that Americans are the most trusting bunch of people in the world, and I suspect it is, then we need to stop and think about what's happening in our society and use some judgment and caution. Caution is definitely the word when it comes to computer or video dating. If you make the decision to go this route—which should be way down at the bottom of your priorities list, a sort of "last-chance cafe"—you ought to give some careful thought to the following:

   1. These services are businesses out to make a profit. No matter how efficient and friendly they seem, they're not doing social work. If you fully understand this point to begin with, you'll save yourself from rude shocks and disappointments.

2. You have as much right, if not more, to examine the service's background as it has to see yours. Insist on getting business references from banks and suppliers, as well as documentable randomly selected case histories and phone numbers of clients from previous years whom you can call for references. If the service refuses to provide you this kind of information, put your checkbook back in your pocket or purse and split.

3. If you're responding to an advertisement placed in a newspaper or magazine, call the advertising department of that newspaper or magazine and ask how long the company has been advertising and whether they have received complaints from readers responding to the ads.

4. When you've satisfied yourself that the service is legitimate, go one step further. When they ask for personal data, give only as much as you feel they'll need to find the kind of person you want to meet. Stay in control. Remember, regardless of what the sales pitch is, everything you give them becomes a part of your file, perhaps to be shown again and again, ''discreetly,'' to strangers.

5. Be particularly careful about using your home or apartment address and phone number. Give them your work address and telephone number, instead.

6. The tricky part comes when you go to meet your match. Obviously, you don't want to start a relationship with an interrogation session. If you're matched up with a corporate type or a working professional, you probably don't have too much to worry about. But your risks skyrocket if you end up with someone who isn't settled and can't easily be pinned down, such as an artist, a student, a writer, an actor or entertainer, a waiter or waitress, or someone who's self-employed. My advice is to spend some time together in a public place before you reveal too much about yourself.

Con Games
(See also Bunco; Elderly, Common Crimes Against the)

Con games are so called because in order to succeed, they've got to get your *con*fidence. They usually have less apparent magic in them than bunco. But you're a mark for either if you believe in getting something for nothing. The major difference between the two is that the operators of con games rely more on your greed and less on tricks and sleight of hand. Unfortunately, not all con games are actionable, or illegal.

KINDS OF CON GAMES

Con games range from out-and-out scams to old-fashioned hustles that have been around for years, disguised as business or financial opportunities. Nearly all of them have the promise that for a relatively small investment, a few hundred or a few thousand dollars, you'll profit many times over and become independently wealthy.

There are almost as many varieties of con games as there are con men. Some operators lure hapless investors into buying phony securities; worthless stock in oil wells, gold and silver mines, and more recently computer companies; or titles to land that's nonexistent, underwater, or hundreds of miles from the nearest road. Others pose as businesses to secure loans. And there are even a few con men who manage to circumvent the law with such finesse that they are able to continue operating their schemes in full view—some of the more questionable franchise schemes, for example. These last are among the most costly and slippery of the con men because they can continue swindling people for years without being identified as illegal operators.

When you buy into a con man's scheme, you only make it more likely that other people will fall into his trap. These operators usually set up the deal so that a small number of you make a few bucks, just enough so that you'll let others know about the quick-return investment you've found. Some of you may smell a rat, but the temptation to say nothing and hang in there in the hope that at least you'll get back your original investment is strong. But if you do smell a rat, you're probably right. And if you're hoping to come out without a loss, you're probably wrong. Your best move is to get on the horn to the state attorney general's office before the con men have time to fold their tents and disappear.

8 WAYS TO AVOID BEING A MARK

1. You don't get something for nothing. This first rule is the hardest. You have to convince yourself that you *never* get something for nothing. That goes against the grain, since most of us are natural optimists. It's easy to go from wishful thinking to the belief that somehow, someday, somebody is going to lay a bunch of money on us through some get-rich scheme. After all, it's really the only way we could ever get rich! That's where we go wrong. No one, but no one, ever got rich from a con game—except the shysters that were running it—and no one ever will.

2. Don't trust strangers. Never give your money or power of attorney to anyone you don't know well and trust implicitly or haven't thoroughly investigated by one or more of the following methods:

- Get the name of the person's bank, attorney, and brokerage. If it's a corporation, check with the secretary of state in the state of incorporation to make sure the company's in good standing and legitimate.

- Run the name past your local better business bureau for possible complaints. The number's in the phone book.

- Ask for references from the people or companies the person does business with: suppliers, printers, the manager of the building where their offices are located, etc.

- Look them up in *Standard & Poors, Dun & Bradstreet*, or other business directories available at your public library.

- Never sign anything until you have thoroughly documented the deal beyond reasonable doubt and have more than satisfied yourself that it's safe and solid.

3. Watch the news. Keep an eye on local news for reports on new con games being run in your area. If anyone approaches you with a proposition that sounds suspicious, call your local police and ask for the bunco squad. You'll not only save your own money, but that of your neighbors as well.

4. Gambling does not pay. Unless you've got money to throw away, don't let yourself be suckered into putting it into lotteries or gambling—regardless of who's running the show. There aren't many games around anymore where you won't get cheated. And even if you're lucky enough to find one that's on the level, the odds of your winning when playing by the house rules are so long that a whole list of things, from being run over by a 1907 Olds to drowning in your shower stall, will probably happen to you first.

5. Hold on to your money. Don't make loans to anyone who offers as collateral unregistered stock, deeds to property you haven't title-searched, untitled motor vehicles, or unauthenticated items, including negotiable securities, artwork, and antiques.

6. Check it out before buying it. Don't buy expensive merchandise, particularly jewelry, precious gems, paintings, or other forms of art from people you don't know and haven't investigated, even, and especially, if they offer them at "discount" prices. Have an expert appraise any merchandise you are considering buying, but there is still the possibility that the merchandise is stolen.

7. Don't open your door to people you don't know. I say this over and over again in this book. It's a cardinal rule of personal security. It will also protect you from a lot of phoney magazine subscription and charity pitches.

8. Check out charities. Before donating money to new or unfamiliar philanthropies, make sure they're listed as approved for tax-exempt status. A list is usually available on request from your state consumer affairs department.

Unlike many other forms of crime, con games, bunco, and scams depend totally on your cooperation. You and your crime-conscious fellow citizens can put an end to them merely by refusing to go along with them. For example, the old shell game has been around in an endless variety of forms since the dawn of greed. You and I may not be the ones to put an end to it, but we can sure put a dent in it if we'll just make up our minds not to be suckers and to follow some of the precautions I suggested.

Credit Cards
(See also Airport Crime; Larceny/ Theft; Pickpockets; Purse Snatchers; Vacation Travel Tips.)

Let's face it, credit cards are practically the same as money today. In most of our regular purchases we use them as easily as we do money. And that's the problem. If we can use them so easily, so can anybody else. Maybe a lot of us would handle them more carefully if we thought of them as cash rather than as plastic. Don't wait until you lose a billfold full of them to take a few extra precautions to keep some thief from using your cards like cash.

8 PRECAUTIONS TO PROTECT YOUR CARDS

1. **Don't collect credit cards.** It's foolish to have more cards than you need or use, however impressive a billfold full of them looks to your friends.

2. **Don't carry them if you aren't going to use them.** Taking them along to your kid's softball game or to the beach, for example, is taking needless risks.

3. **Stash them at home.** Set up a card file in a secure place in your home; for example, in a desk that can be locked, a steel cabinet, strongbox, or safe. Get in the habit of carrying only those cards you intend to use. Leave the others safely locked away, just as you would large amounts of cash.

4. **Carry them safely.** Avoid carrying cards in an open handbag or in your hip pocket where they're easy targets for pickpockets. Carry them in purses with zipper or flap closures

or in inside jacket pockets. If you don't wear jackets, at least make sure you keep your hip pocket buttoned when your card-filled wallet is in it.

5. Keep a list. Make at least two copies of a list of all your credit card account numbers together with phone numbers to be called in case of loss. Keep one copy with your papers at home and one at work. If you rent a safety deposit box, keep an additional copy in it.

6. Take an inventory. When you're traveling or regularly using your cards, take a daily inventory just before retiring to make sure none has been lost or stolen.

7. Check your card after using it. When eating out, don't walk out without the card you've just used to pay for dinner. And double check to make sure the card you get back has your name on it. Waiters and cashiers are busy people, and it's not unusual for them to get confused and return the wrong card.

8. Rip up those carbons. A recent wrinkle is criminal use of the carbon papers from your charge vouchers to obtain "good" charge account numbers for use in counterfeiting cards or making illegal telephone purchases. It's getting so you can't even leave your trash behind anymore. To be safe, after you've signed the charge voucher, remove the carbons, rip them up and throw them away yourself.

WHAT TO DO IF YOUR CREDIT CARDS ARE LOST OR STOLEN

If you've read the promotion literature you find enclosed with your charge card bill, you've probably read all this before, but it's important and bears repeating.

Most states have laws limiting your liability for misuse of your credit cards to $50 per card. That's reassuring; but if you were carrying 10 cards and all were used by thieves to make purchases, you'd still be looking at *five big ones*.

It's possible to insure your cards against loss or theft. But the rider on your regular home or apartment insurance covering out-of-home losses may already cover you for charge card losses. If not, call around to insurance companies until you find one that does cover charge card losses.

In the event you *do* lose one or more of your cards, it's important that you report the loss quickly:

1. Use your list of cards and numbers to report their loss to the issuing companies. Do this immediately upon discovering your loss *and confirm the calls in writing.* Make two copies of your letter in case any question comes up as to when you reported the loss.

2. Notify your local sheriff or police. Give them a list of card numbers and a written statement of how the loss occurred.

3. If you have insurance on your home or apartment that includes out-of-home loss coverage or other insurance covering your cards, notify the company both by telephone and in writing at the same time you notify the card companies.

D

Doors
(See also Alarms, Home and Business; Apartment Security; Burglary; House Security; Locks; Personal Safety.)

You may be one of the lucky ones who lives in a house or apartment with good, "hard" metal or metal-sheathed wooden doors, either because you had the foresight to have them built in or because they were there when you moved in. Most of us don't. Most of us rely on flimsy doors and door frames equipped with the most rudimentary locks for our security. The fact is, a determined intruder can get through these doors in seconds simply by punching a hole through them with his fist and opening them from the inside or by kicking them open with his foot.

KINDS OF DOORS

Before we talk about hardening your doors, you should know what to look for. Although there are many different designs, shapes, sizes, etc., doors can be broken down into four basic types:

1. Sheet metal (usually steel) on metal frames. These

doors offer maximum security, both from intrusion and fire. They're usually made from sheets of 14-to-18 gauge steel formed and welded to a metal frame, providing both lightness and strength.

2.   Solid wood or solid-core plywood (sometimes sheathed in sheet steel). These rank second in security but are less fire resistant.

3.   Hollow-core plywood. These doors, common in both apartments and houses today, are best described as sheets of plywood glued to light wooden frames. Except for an occasional stiffener that crisscrosses the frame to minimize warp, these doors are hollow. There's nothing but air between the two thin sheets of plywood.

4.   Glass. These doors are generally the least secure, whether the glass is framed in wood, as in the French style, or framed in metal, as in modern sliding doors.

Testing Your Doors and Frames

You can determine what kinds of doors you have by performing a few quick tests. While performing the tests, keep in mind that the best doors in the world are useless if they're attached to weak or flimsy jambs or frames or secured by cheap, bevel locks. Good doors should be set in good solid frames and have good quality, dead-bolt locks.

1.   Testing painted doors. If the door is painted, scratch away a small amount of paint to determine if it is wood or metal. The same test can be made on a painted door frame. Glass doors, of course, are easily identified.

2.   Testing wooden doors. With practice you'll be able to distinguish between the solid and hollow-core door simply by tapping lightly with your knuckle or the handle of a screwdriver or other tool. The solid door will give a deep, firm sound, whereas the hollow-core door will give a higher, thinner sound. If you're still not sure, try driving a thin finishing nail into the plywood. If it goes in a bit but continues to resist your hammer after a few blows, the door is probably solid. If it suddenly yields after your first few hammer blows, the door's hollow and should be replaced. There's no sense in putting a $50 lock on a $10 door.

You may need to test more than one spot on the door as hollow-core doors have the above-mentioned stiffeners in various locations beneath the plywood. You can plug the resulting nail holes with small amounts of filler or plastic wood and paint or restain them.

3. Testing prefabricated doors. There's a growing trend toward prefabricated metal doors and frames. Unfortunately, many homes still have lightweight frames—half-inch wooden strips wedged and nailed with six-penny nails to the door openings. If your doors are hollow core, chances are the frames are of these half-inch wood strips trimmed with light moldings where the door opening meets the walls. These will not provide you with good security and should be replaced.

HARDENING YOUR DOORS

There are two good arguments for hardening your doors:

1. It makes it difficult, if not impossible, for amateur burglars, particularly the kids responsible for a disproportionate number of burglaries, to break in *while you're out*.

2. It increases your own and your family's security *while you're at home*.

Door Frames

Again, you may be one of the lucky ones who already have hard doors. If not, here are a few suggestions.

Solid Wood Doors. These include solid-core plywood doors and metal-sheathed wood doors. These doors are fine, but make sure the jamb and frame are equally strong. If the door stop (the strip of wood that runs the length of the door frame that stops the door when closed) is on the outside, make sure it's firmly fastened with recessed nails—or better yet, slender screws—to resist being pried loose. Check hinge and strike plate screws to make sure they're long enough to anchor firmly in door or frame joist (see p. 95, items 1–4).

Hollow-Core Doors. If you have hollow-core entry doors (front, rear, or garage door), make every effort to replace them. Measure the doors carefully, including height, width, and thickness. Then take the measurements to your local building supply or lumber yard and get advice on the best door for your security needs. Usually the next step up, from hollow to solid-core plywood doors, will serve your needs at reasonable cost.

Glass Doors. These are a security nightmare—particularly glass doors in wooden frames that are usually located off patios, terraces, or lower balconies. They're as risky as they are elegant. The most effective way to protect them is with interior steel gates that can be folded away out of sight when not locked. These are expensive, difficult to conceal, and tend to defeat the purpose of the decorative doors. As I've mentioned before, there's also the problem of safety from fire. Unless everyone likely to be in the house knows how the gates work and how to unlock them—which presupposes the key must be left someplace accessible—there's the risk of being trapped in a smokey room.

Glass doors in metal frames are less of a problem, since these tend to be larger sliding glass doors with double-thick, tempered glass. They can still be broken by a determined intruder. Usually one large section is permanently attached to the building in an aluminum frame while the other section slides open and closed on tracks set in the upper and lower frame. The sliding portion is usually lockable but with the lock secured by a lightweight latch that can be easily broken.

Another problem is that unless these sliding glass doors are equipped with stops after they're installed, they can be easily lifted out of their tracks. If you have sliding glass doors that have never been equipped with these safety stops, discuss your problem with your local hardware person. There are a number of reasonably priced security devices on the market designed to protect these very popular glass doors. As a temporary measure, you might try inserting a wooden dowel between the top track and the top of the glass door frame. The dowel should be of appropriate size to slide loosely into

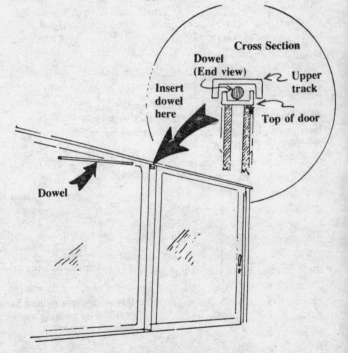

Figure 7. SLIDING GLASS DOORS.

the space between the tracks of the door and frame (see Figure 7).

Another technique for securing sliding glass doors is to drill holes in the outside flange of the top track of the door frame and insert screws in several places to prevent lifting the door past them. In the section on ''Locks,'' I also mention a hydraulic closer and locking mechanism which can be purchased for these doors.

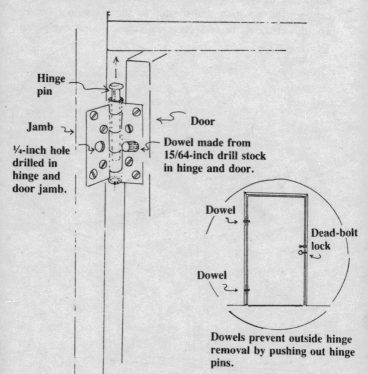

Hinge pin

Jamb

¼-inch hole drilled in hinge and door jamb.

Door

Dowel made from 15/64-inch drill stock in hinge and door.

Dowel

Dead-bolt lock

Dowel

Dowels prevent outside hinge removal by pushing out hinge pins.

Figure 8.   PREVENTING DOOR REMOVAL FROM HINGES.

Hinges

At the same time you're inspecting your door frames, take a look at the hinges:

1. Are they large and sturdy?

2. Are the tubular sections that rotate on the hinge pins set on the inside (or inner "secure" side) of the door?

3. If for some structural reason, hinges *must* be located on the *outside*, are the hinge pins fully driven into the inter-

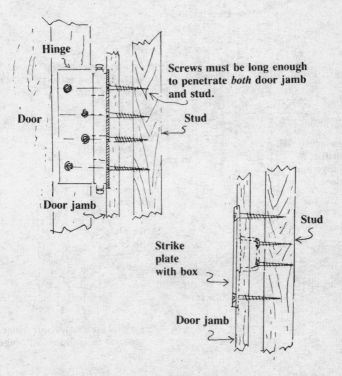

Figure 9. HINGE AND STRIKE PLATE INSTALLATION: THE RIGHT WAY.

locking hinge sections and well secured in some way? One way to secure a hinge pin is to flatten or enlarge the end of the hinge pin by striking it several times with a hammer while holding a metal wedge or other suitable backup to the hinge pin head to prevent it from moving—an operation known as peening. The purpose of peening the hinge pin is to prevent anyone from simply pushing the hinge pin out and removing the door. One further means of securing doors from removal—

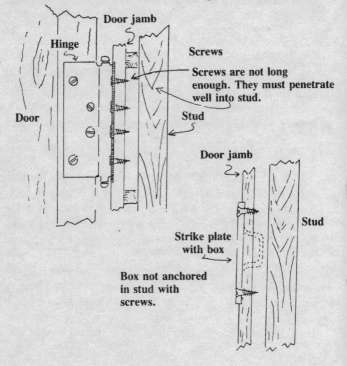

Door jamb

Hinge

Screws

Screws are not long enough. They must penetrate well into stud.

Door

Stud

Door jamb

Stud

Strike plate with box

Box not anchored in stud with screws.

Figure 10. HINGE AND STRIKE PLATE INSTALLATION: THE WRONG WAY.

even if hinges *must* be located on the outside of the door—is through use of steel dowls through hinge plates, as illustrated in Figure 8.

4. If the hinges on your solid wood door are good-sized and attached on the inside to a strong wooden or metal frame, remove one of the hinge attaching screws with a screwdriver. Is it at least 2 inches long or long enough to securely attach the hinge to the stud of the wall opening as well as the jamb? If not, replace *all* the hinge-attaching screws (see Figures 9 and 10).

Alarms

As I point out in the section on "Alarms, Home and Business," all doors, as well as windows and other access points, should be secured by some form of alarm.

Hard doors and door frames are important elements in your own private war on crime, and a good alarm system is an essential backup. Good crime prevention requires both.

If you're reading this at the same time that you're thinking about buying a new home or condo, you're doubly in luck. Security is a lot easier to come by when you can build it into new construction, particularly during the ground-breaking stage. Now's the time to remind the builder of your bottom-line requirements for security.

In designing your custom home, chances are your architect has already provided for the increased security demanded by the times. But if you plan to buy in a development, you'll have to specify any extras well in advance of construction. Hard doors and door frames should be at the top of your list.

Drugs in the First Degree
(See also Burglary; Drugs in the Second Degree; Kids; Larceny/Theft; Mugging; Murder; Personal Safety.)

The drugs I talk about in this section are the hard stuff: heroin, morphine, and opium. Heroin's the main drug on the streets, where it's usually called "smack." It's also the most addictive and destructive. Cocaine is also officially classed as a "hard" drug, but it's milder than the others and is similar to the stuff your dentist gives you as a painkiller. Like marijuana, cocaine isn't addictive physiologically—but it builds psychological dependence. Because the usual way of taking it is by sniffing, prolonged use damages tissues in the nose and throat and can lead to serious infections and irreversible physical damages.

MYTHS AND THE REALITY

I'll tell you the truth about drugs so you can protect your own life and the lives of others from their destructive effects. Drugs are every bit as dangerous, physically and socially, as we've been told they are. Unfortunately, we've built up a bunch of myths about them that tend to perpetuate their use. And that's not only by thrill seekers and criminals, but by otherwise law-abiding citizens and, worst of all, by our kids. I used heroin for over a year on a daily basis, and before I quit I was spending $50 to $60 a day. At the time that amounted to two or three "spoons" a week, which is a lot of "smack." In that year, I learned a lot about people and drugs, and the myths they use to justify drug use.

Drugs in the First Degree

MYTH—*People who use drugs are part of a very select and mysterious group known collectively as the "drug culture."*

REALITY—There's nothing "cultured" about drug addicts, regardless of what they were like before getting into drugs. Drug addiction is an avoidable, aberrational, clinical disorder of deadbeats. It is antisocial—not social—behavior.

MYTH—*It's "smart" or "chic" to use drugs, particularly cocaine, which many top celebs, jocks, entertainment and social register types use.*

REALITY—When I was a kid in California, there was a saying on the street, "Hard stuff is *the dumbest*." The people you heard say it most often were jazz musicians, among them my father, who regularly smoked marijuana with other musicians. At the time, he was part of a tiny minority of artists who'd been smoking "gage" or tea," as they called it then, for many years. Yet, to the man, those cats agreed that using hard drugs was not cool: It was stupid and self-destructive, and something they wanted no part of. That's still true. Smoking pot is not smart—but hard stuff is still the dumbest.

MYTH—*You haven't lived until you've tried hard stuff. Hard drugs give you a "kick" you get nowhere else.*

REALITY—As an addict, I spent 10% of the time stoned out of my gourd feeling just great; the other 90% of the time I spent trying to keep from feeling lousy. For most of the year that I was into drugs, I was into maintenance, period! You keep needing more, and the kick you get isn't worth all the hassle and all the creeps drugs put you into bed with. You become a victim. Drugs are the dregs.

MYTH—*We're helpless victims of the drug pushers. The mafia* *is responsible for bringing most of the drugs into this country, and it's up to our government to stop them.*

REALITY—You might as well say the Russians, the Ayatollah, the Pakistanis, or the Communist Chinese are responsible. You'd probably be just as right. As long as *we*, you and I, create a demand and a profit can be made, whether economic or political, there will be people willing to fill the demand. And profit is as important to governments as it is to criminals. It's up to us to get our heads straightened out and to stop creating a demand. In my opinion, that's the only way it's ever going to stop.

MYTH—*We need stiffer laws, stricter judges, and better enforcement, to eliminate drug sales.*

REALITY—As with other crime, stiffer laws and penalties are no deterrent. Criminals don't really believe they'll get caught. If they did, they wouldn't do it. New York has some of the toughest narcotics laws in the country—and one of the worst drug problems. I don't care how many laws we pass or how tough the judges get, as long as people buy the stuff, there'll be someone there to sell it to them.

KICKING THE HABIT

As is often the case, I got into heroin through an acquaintance, a musician friend who was an addict. Some friend! But that's the way it usually starts. He introduced me to other users and their suppliers around town. By the end of the first year, I was pretty well hooked.

It was spring in San Diego, I was a cocky 18-year-old hood, and I was in love with a beautiful girl. She was a cheerleader at a local high school, but she was old for her age. At the time we connected, I'd just made the biggest score of my life. I had plenty of money for other things, as well as dope. I'd just bought a shiny new convertible with white sidewall tires. I was taking this luscious California girl to a Kenton concert in Balboa. Front row seats. We were going to arrive and walk down the aisle, and all the local dudes and their dates were going to see how sharp and cool we were. It was to be the major social coup of my life. I went to the finest men's store in town and bought a new set of threads, including a gold shirt. In the shoe department, I bought a pair of white suede loafers that I had dyed gold to match the shirt.

The afternoon before the big date, I realized I was out of dope. From there on, it was downhill. The guy I usually bought from had gotten busted or was out of town. Some street friends gave me the name of a pusher known as Mosquito. Mosquito turned out to be a creepy-looking Mexican dude who, when we established that I wasn't a cop, got into my car and directed me down a muddy alley to the rear of a dilapidated old house. Mosquito told me to wait on the porch while he got the stuff.

I stepped out of my car, and the next thing I knew I was up to my ankles in black mud. The gold loafers looked like they belonged to the referee at a Texas hog call. I got to the porch and stood there, waiting for Mosquito to bring the "smack." I looked at my muddy shoes. I had my money in my hand, money I'd risked my life for. The scene filled me with disgust. I thought, "Man, I am insane! Only crazy people would do something like this for anything as stupid as dope!" I got into my car and went home.

I haven't touched narcotics since. Kicking the habit was unpleasant. But from my own experience and from working with addicts at rehab facilities since, I've found that most addicts can kick a habit without much worse symptoms than those associated with a bad case of flu—aches, pains, chills, sweating, sometimes vomiting and diarrhea. It's not something you'll want to do every day, but you've got to believe it's one hell of a lot better for you than an overdose. I never knew anyone who died from going cold turkey.

HOW DRUG CRIMES AFFECT YOU

Most of us know somebody who's into drugs in one way or another. For a lot of different reasons, we often look the other way. We've become believers of the myths—and the cost is staggering. *We're spending over $200 billion a year on drugs*, roughly the same amount we spend on national defense.

I don't think there's anything more important than getting on top of this drug problem. It's a more pressing problem than unemployment, foreign policy, or other crime. Over a half-million people were arrested for drug offenses in 1981, and estimates of the number of users of various forms of narcotics, excluding alcohol, run as high as 5% of the population, including large numbers of kids. Although optimists claim drug usage is down, according to the Uniform Crime Report, "In the period 1978 to 1982, arrests for these violations increased less than 1%."

The only way to stop the drug traffic is to stop looking the other way. When you do, you're not doing anybody, not yourself or anyone else, any favors. If it's someone you're close to, try to talk to that person. But if turning the person in is what it takes, you'll have to accept that. You're not violating any code by saving someone's life—and that's really what you're doing when you try to get someone off drugs.

HOW TO ACT WHEN CONFRONTED BY AN ADDICT

I recently saw a survey of victims of armed robbery. One of the questions asked was "Did your robber appear drunk or drugged?" Fewer than 1% of respondents answered yes.

Violent crimes, such as armed robbery and assault, are too nerve-racking and risky for most addicts. Although addicts get the blame for a lot of mugging (which is really armed robbery) and push-ins (which is assault and robbery), the largest number of these crimes are not committed by addicts at all, but are primarily the work of violent street criminals. Addicts prefer to con money from their friends and relatives. In a pinch, they'll turn to larceny/theft, which includes purse-snatching, dipping (picking pockets), burglary, and prostitution—crimes they are 20 times more likely to commit than violent crimes.

These are the crimes in which you're most likely to confront an addict. And unless he or she is "crashing" (suffering withdrawal symptoms) or you happen to be familiar with the signs of addiction, you may not even realize your assailant is an addict. Here are a few tips in case you do find yourself fact to face with an addict:

1. **Avoid confrontations.** Avoid contact with any suspect at the scene of a crime, particularly burglary or theft, where chances of confronting an addict are greatest. Stay out of sight if you can. When it's safe, call the police.

2. **Don't panic.** If you can't avoid a confrontation, try not to panic or show fear—keep calm. Many addicts turn to crime to get money for their habit only after they begin to crash and are desperate for more drugs. In this condition they're edgy and easily upset. Although addiction is actually less of a factor in violent crime than popularly believed, the fact remains that many of the people who take drugs are

dangerous—on or off drugs. As I advised you in "Armed Robbery" and other sections, don't lecture, lose your temper or get hysterical. Follow his instructions, give him what he wants, and get rid of him.

3. **Use your common sense.** If anyone you suspect of being an addict confronts you in the street or other public place and demands your purse, use your common sense. Although it's rare, petty criminals like purse snatchers and pickpockets have been known to hurt people. My advice is to give the creep your purse if he's got a hand on you or is close enough for contact. (See also "Purse Snatchers.")

Drugs in the Second Degree
(See also Burglary; Drugs in the First Degree; Kids; Larceny/Theft; Personal Safety.)

ALCOHOL

Alcohol is the most widely used narcotic in the U.S.—and in fact, in the rest of the world. In this country, it's the only major acknowledged narcotic and controlled substance sold without a prescription. It's considered to be only mildly addictive but it can be extremely destructive for some people. Like other drugs, it can be lethal if misused. Because of its broad availability, general use, and acceptance in the eyes of society, more crime is committed by people under the influence of alcohol than any other drug. Murder is a prime example. It's all too often the case that murder is committed by someone known to the victim and that the murderer is under the influence of alcohol at the time of the murder.

DRIVING WHILE INTOXICATED—
THE NO. 1 VIOLENT CRIME

Many Americans think nothing of mixing drinking and driving. They've been doing it ever since Henry pushed his first Tin Lizzie out of the barn. In those days, there wasn't much of a chance of anyone getting hurt. There weren't many cars on the roads, and you couldn't keep them running long enough to do much damage.

Today, there are approximately 124 million cars crowding our streets and freeways, and over 50,000 vehicular deaths occur each year, more than twice the number of people murdered in violent crimes. Although many of these deaths happen by accident or out of driver carelessness, between one-third and one-half are known to be the direct result of drunk driving. In other words, almost as many people die each year as a result of someone's driving while intoxicated as are killed in homicides. And that's just for openers. Of the more than 5 million people injured annually in auto accidents, as many as 2.5 million result from drunk driving.

The loss of lives and the cost in human suffering, time, money, and productivity is beyond counting.

As in the case of other drug use, only we can put an end to it. No amount of legislation, no degree of punishment or penalty is going to stop alcohol abuse.

Keeping Drunk Drivers Off the Road

Putting an end to drunk driving, like charity, begins at home. First, rearrange your life so you won't find yourself behind the wheel when you've had a few belts too many. After all, you're the *last* guy who wants to end up losing his driver's license—or worse. Here are some tips you should follow:

1. **Take a taxi.** If you find yourself spending the after-

noon in a bar, leave your car in the parking lot and take a cab home.

2. Let the other guy win. If you're a sales person unexpectedly locked in with a client who turns out to be a big-league lush who insists on a drinking duel, let him win. Then grant him the privilege of paying your cab fare home and your overnight parking tab.

3. Report that driver. If you spot a car that's weaving all over the road, don't just ease your car around him and drive on, shaking your head. Get on your CB or stop at the first pay phone and call the police or highway patrol. Many cities have special numbers to call when you spot a drunk driver. If your area has one, write it in your address book for future reference. You can call your operator to get it. Try to give the police a full description of the vehicle: make, color, license number, and speed and direction of travel (nearest milepost if on the highway). If turning a drunk driver in makes you feel like a rat, just remember that you could be saving his life, as well as the lives of innocent motorists and pedestrians.

POT

Marijuana, more familiarly "pot" or "grass," is the most common narcotic used in the U.S. after alcohol. Kids usually get into pot because of peer pressure. The growers and distributors make sure it's plentiful and cheap in that market—at least in comparison with what they sell to the same kids later on. The greater danger of pot is that those who are attracted to it in the first place often go on to stronger stuff.

There's a lot of controversy about whether pot is as bad for us as alcohol. Most of the arguments are old hat by now. Everyone's pretty well made up his or her own mind about it, so I try not to get into it. The one thing I do tell people is simply this: There's a lot of pressure on kids to try pot. If your kids are hanging out smoking pot with other kids, sooner or later, someone is going to say to them, "If you think *that's* cool, wait until you try some of *this!*" And you, and they, better be ready, because if they buy that line, chances are you're going to lose them. They may find their way back somehow, like I did—25 years later. But boy, I want to tell you! You'll both go through a whole lot of pain and heartache in the meantime!

Is Your Kid Using Pot?

How can you tell if your teenager is into pot? Here are some clues:

- Has your kid's personality changed?

- Have school grades fallen off suddenly?

- Is there a lot of "dating" and "meeting" going on?

- Is your formerly well-adjusted offspring suddenly vague, distant, and snappish by turns?

- Does he or she have the "munchies," constantly nibbling and drinking unusually large quantities of Coke, Pepsi, and even water?

- Have colored lights, blinking strobes, strange music, and the smell of incense taken over the fledgling's nest?

- Does your kid claim that the recently increased allowance is woefully inadequate to the needs of the day?

- Have you begun to wonder why you keep smelling burning straw in the apartment, house or yard—or when you hitch a ride in your kid's car?

Adolescence isn't an easy time under the best of circumstances, and several of these symptoms might be normal for certain kids. But if you recognize more than a few of them, particularly the last two or three, get some help. It's natural to be frightened by the thought that your child is smoking pot, but don't let your fear prevent you from talking to your kid and trying to alert him or her to the potential dangers of drug abuse.

Don't let anyone tell you that marijuana isn't a narcotic and that smoking it is harmless. It's true that some kids get through it without hurting themselves, but too many don't.

Check with local school and religious groups for services to help you. Most communities have narcotic guidance councils or narcotics anonymous chapters; they're listed in the phone book. There's no charge for these services, and experienced people will be happy to give you practical advice on how to deal with your kid's involvement with drugs.

E

Elderly, Common Crimes Against the
(See also Bunco; Con Games; Mugging; Personal Safety; Purse Snatchers; Push-Ins; Social Security Theft.)

A whole lot of the crime against older people happens in our big cities and surrounding metro areas. What I've got to recommend is going to be a hardship, but the fact is, if you're over 65 the surest way to prevent crime from happening to you—or at least to reduce your exposure to it—is to go someplace where there's less of it. I urge all the elderly I talk to to consider getting together with other elderly people, relatives, or friends—whatever it takes to get out of the metro areas into smaller, safer communities.

The potential for crime against the elderly in big cities is well known. If you're an older person living in a city, you know the list better than I do: Social Security check theft probably heads it. If you're living on small savings and retirement benefits, you're a mark for bunco and confidence schemes that tempt you with the promise of additional income. If your retirement income is limited and you live in a low-rent city neighborhood where punks and addicts tend to hang out, street crimes like push-ins, purse snatching, mugging, and assault are a serious threat to your safety.

If you can't move to a safer community, here are some things you can do to lower your risk of becoming a victim where you do live.

8 WAYS TO REDUCE YOUR EXPOSURE

1. Add some security to your Social Security. Don't wait until your check is stolen out of the mailbox or you're repeatedly robbed or mugged. Arrange with your local bank for direct deposit, and your Social Security check will be mailed directly to the bank and credited to your account. You can then write checks against it without ever exposing your check or the cash to robbery. When the neighborhood thugs get the word that you never carry cash, they'll leave you alone.

2. Don't trust strangers. Avoid financial dealings with people you don't know or have only recently met. If they are overly friendly, be especially cautious. Don't give them any money and by *no* means sign over your power of attorney. Don't sign anything without consulting someone you know and trust. Talk to friends, family, or a counselor at your local senior citizen service center, whose telephone number and address you can find in your phone book.

3. Be alert to push-ins. Push-ins of elderly citizens have become the stock-in-trade of many of the crooks lurking about our cities. There are two kinds of push-ins. In the original version, the criminal lurks behind you until you unlock your door. Then at the precise moment that you start to enter, he leaps on you and pushes you inside, where you're at his mercy. In the second kind, which occurs when you're already inside your home, you make the mistake of opening your door to a stranger, at which point he pushes his way in.

In either case, your best defense is to be aware of the problem, to stay alert, and to practice good personal security. When you're going into your apartment, be sure no one's lurking behind you. When you're inside, don't open the door to strangers. This includes any solicitors and fund raisers that come to your door. Just explain firmly through the door that you don't do business with door-to-door people, and stick to it.

4. Watch out for con artists. Grieving spouses are likely targets of con game operators who use the obits to put together their "prospect" lists. Watch out for anyone who approaches you with goods he claims were ordered for you by the deceased before his or her passing or who claim money is owed for retirement plans, insurance, religious works, inspirational books, or art objects. Report any such attempts to the police or your local senior citizens service center.

5. Don't go it alone. If you *must* live in the city and travel its streets, don't do it alone. Join a senior citizens group or a local church or synagogue. Sit down with the people you meet there and work out shopping, visiting, and recreation schedules so that none of you will be on the street alone. Try to work out the schedules so that you can get your chores and business taken care of in the morning hours, while the hoodlums are still in bed. By far the largest number of muggings and assaults on the elderly are the work of punks looking for lone, elderly (and easy) targets.

6. Carry a noise alarm. Anytime you're out, whether alone or not, keep handy one of the small personal-security noise alarms described in the "Personal Safety" section. These little pocket alarms, powered by either batteries or aerosol, can be heard up to a mile away and will blow the mold off a mugger's mind at 50 paces.

If you're taken by surprise and unable to use your alarm, don't panic. Remain calm. Most elderly people hurt in muggings become victims of violence when they try to resist or become confused and excited. Usually the mugger is only interested in getting your money. Give it to him and get rid of him. Your life is worth far more than anything else he can possibly take from you.

7. Report the crime. Don't be afraid to report crimes or threats of crimes against you or your neighbors. Criminals who pick on the elderly have got to be cowards at heart. They're big on threats but short on follow-through. Many cities are legislating strict new penalties for crimes against the elderly. But it's up to you and other older citizens to report violations to the police if that legislation is to do any good.

8. Get protection. If you're concerned about the increase in crimes against the elderly, write the director of your local senior citizens organization. He or she can put you in contact with your local representative, law enforcement agency, and other political and governmental agencies whose responsibility it is to protect your rights in the community. Another good idea is to organize an oversight committee made up of members of your senior citizens group. You can also look for help from your church or synagogue, the local police, local crime watch, and block associations.

F

Freeway Crime Smarts
(See also Armed Robbery; Personal Safety; Vacation Travel Tips.)

Imagine the following scene:

You're on a freeway at rush hour. Traffic around you is bumper to bumper. A line of cars is merging from the right, barely moving, squeezing into your lane of traffic. The guy in the car at the head of the line has decided he's getting in, no matter what. You know that you have the right of way, and the driver behind you knows it, even as he helps you down the freeway with his front bumper.

The guy at the head of the line keeps coming. He's determined. His eyes are wild and he's mumbling to himself. Then, like two ships coming together in the dead of night, your two cars collide in a slow grind. The other driver erupts from his car like a scalded cat and dances across your hood screaming at the top of his voice. You lucked out if he wasn't carrying a gun in his car.

Every so often you hear about some dude who climbs out of his car with a gun and starts shooting. People are getting killed over minor traffic arguments. It's war out there, and it doesn't make any sense.

TIPS FOR THE FREEWAY

All of us who drive need to step back and take a look at what we're doing. Here are a few suggestions that may help you keep from becoming a statistic:

1. Take it easy. When you're out on the freeway, force yourself to take it a little easier. Give everyone a little more room. You'll be surprised how much your stomach will appreciate your new driving style and how much more you'll enjoy the trip.

2. Give yourself time. Make it a point to leave for appointments early. Allow more time than you really need. Don't get caught in the freeway time warp on your way to an important appointment.

3. Let the other guy get ahead. When an aggressive driver wants to butt in ahead of you, let him. You'll probably never see him or his car again. Don't try to show him how much faster your Porsche will go than his Ferrari, or box him in while you and the driver of the car next to you dawdle along together at 55. And there's always the chance the guy in the car trying to get around you is legitimately in a hurry—an undercover cop chasing a drug dealer or someone about to give birth.

4. Don't risk provoking the wrath of your fellow drivers. Fight down the temptation to tell the guy behind you where to put his horn. Avoid loose talk or arguments with strange drivers. Don't gesture or taunt other drivers. You don't know who you're talking to. He may be a pussycat, or he *may* be Atilla the Hun. It's a gambler's choice and, frankly, considering some of the people I met in the joint, I just don't like the odds.

5. Ignore the creep. If another driver gives you a hard time, ignore him. Don't stop and get out of your car with the idea of giving him a punch in the nose. He may have a whole different idea of how to resolve the argument. If he persists, get on your CB and call for help. If you don't have a CB, keep driving till you lose him or see a cop you can flag down.

6. Be alert to bump and rob operators. This is particularly true late at night. It usually happens like this: There will be two guys in a beat-up old car who run into the back or side of your car. When you stop to check the damage and exchange insurance information, they rob you, pocket your money, valuables, and keys, and drive off. For tips on prevention, see page 40.

7. Act smart in time of trouble. If you have mechanical trouble or a flat tire on the freeway, do whatever is necessary to get your car off the road and well out of the traffic lanes. If you have flares or safety reflectors in your emergency kit, put them 200 feet or so down the road from your car. Then go back to your car, lift the hood, and lock yourself inside. Use your CB to call for assistance, or wait for the freeway police or service vehicle to find you. Don't unlock your car door or roll windows down to talk to strangers. If the driver of another vehicle stops and offers help, roll your car window down an inch or so and ask him to contact highway police or emergency road service from the next pay phone.

8. Don't carry weapons. Don't carry guns, chains, tire irons, taped pipes, shot bags, saps, billy clubs, knives, or other weapons for self-defense in your car. Weapons usually end up hurting somebody or getting somebody hurt. And regardless of how you come out of the fight, you can be the other victim whose life gets messed up; that is, you could end up being jailed for assault or even worse.

The botton line is if you exercise good sense, show ordinary courtesy, and obey the law, you'll probably never need to worry about freeway crime. You'll contribute to its prevention, and you'll feel better about yourself.

G

Guns
(See also Murder; Personal Safety; Self-Defense.)

SHOULD YOU OWN A GUN?

I'm against people owning guns, especially handguns. No one knows better than I do that there are some really bad characters out there with guns in their hands who won't think twice about using them. But the truth is, most crime is about money. The average crook isn't out to get you; he just wants your money or whatever you have that he can sell for money. Most crooks won't hurt you unless you do something dumb to provoke them or put them in jeopardy—like pulling a gun.

There are exceptions: Maybe one out of every 2,000 or 3,000 offenses is committed by a psycho. But your chances of running into one of *them* are roughly the same as your chances of making five straight passes in Las Vegas. Those odds may be good enough for you to justify carrying a loaded gun and a pocket full of bullets for the rest of your life, but they're not for me. To me, guns are an inconvenience that interfere with my comfort—as well as a danger.

So your decision about whether or not to buy a gun should be given a lot of thought. Here are some questions to ask yourself:

1. How much do you really know about guns?

2. How adept are you? Could you handle a gun smoothly, without getting it all hung up on your belt loop or purse strap?

3. Do you consistently hit what you aim at?

4. Do you know how to clean a gun and take care of it?

5. Do you know ammunitions and understand the differences among various kinds?

6. Will you be able to keep it out of the wrong hands— your children, teenagers, burglars?

7. Are you prepared to shoot to kill to prevent someone from killing you?

Reasons Not To Own a Gun

One statistic should haunt us all. There are more accidental deaths and suicides by guns than there are intentional homicides. The point is that people are just not careful enough with their guns.

Unless you belong to a gun club and shoot every week, you probably don't have much chance to practice shooting and handling even the ones you do own. Meanwhile, many of the robbers and other crooks who use guns in their trades are gun nuts. The guns become extensions of their personalities. They're always doing something with their guns: cleaning them, handling them, shooting them at public ranges or out in the sticks somewhere, or reading about guns and ammo in magazines.

Guns are part of these guys' business. And unless you're just as serious about guns as they are, you're a whole lot better off unarmed, taking your chances with the rest of us.

H

Hostage Situations
(See also Armed Robbery;
Kidnapping; Personal Safety.)

Compared to your chances of being a victim of theft, burglary, or robbery, your chances of facing a hostage situation are slim. Being taken hostage is an experience most people get through life without. But hostage situations are on the increase, and they're occurring more often in public places. So you should be prepared in case it happens to you.

TYPES OF HOSTAGE SITUATIONS

There are three basic kinds of hostage situations:
1. Hostages of street crimes, where the person is taken for ransom or to ensure escape, as in a foiled holdup or airline hijacking.
2. Hostages taken during home robberies, usually to obtain ransom.
3. Political hostages, usually taken to effect an exchange of political prisoners or for some political goal or to make a statement.

The most common hostage-taking situation in the U.S. occurrs in the progress of a robbery, particularly when a robbery attempt goes sour. The police interrupt a holdup and the robbers grab you and some other people, blockade the entrance, and trot you out as their "insurance."

Another potentially dangerous situation is the unexpected minivacation several thousand Americans have taken to Castro's island, and other depressing places, courtesy of airplane hijackers. Many of these would-be hijackers turn out to deranged turkeys carrying an empty bottle of Thunderbird in a paper bag disguised as a bomb. They're basically harmless. But the cost of finding out which one isn't is high.

The less common situation, at least in this country, occurs when a religious or political group makes its move to take hostages for ideological reasons. This is more likely to happen if you're visiting in Europe or the Middle East or if you're employed outside the U.S.

HOW TO HANDLE A HOSTAGE SITUATION

The most important thing to do in a hostage situation is stay calm. Don't get excited. I'll share a story with you to show you what I mean.

> Back in the fifties when I was doing a lot of robbing, one of my gigs involved holding supermarket employees hostage. I went in solo just before closing time and pushed a cart of groceries around until the other shoppers left. Then, one by one, I showed the half-dozen employees my gun and escorted them back to a meat locker.
>
> When I finally got to the manager and down to business, I discovered that the store safe was kept in a room with the biggest plate glass window in the city of L.A., with people and cars going past it like the Easter Parade.

My only choice, if I didn't want to go away empty handed, was to send the manager up to the safe alone. I gave him the usual speech about the lives of the hostages in the meat locker being in his hands (there's a lot of show biz in armed robbery) and sent him after the the money in the safe.

The whole thing went off without a hitch. The hostages were staying calm, and the manager played his part perfectly and brought back three money bags from the safe. I checked the first one he handed me—it was full of pretty green stuff. He joined the others in the meat locker. I closed the locker door and got out. A few blocks away, I stopped at a pay phone and called the police to tell them to get the employees out. In fact, I made a couple of calls to different precincts, just to make sure they rescued those shivering hostages. Hell, I didn't have anything against the employees. They didn't give me a rough time.

When I got back to my pad, I dumped the contents of the first bag and counted it. Then I dumped the second and the third. I couldn't believe it! My bed was covered with piles of manufacturer's redemption coupons cut out of magazines and newspapers! There wasn't a dime in either bag!

The next day I called the manager at the store I'd hit. I told him I admired him and his employees for being so cool, but that he was sure one dumb sunavabitch! I said I'd known guys in the joint who, if they'd opened one of those bags stuffed with coupons, would have just blown him away.

If you find yourself taken as a hostage during a crime, here are a few tips for getting out of it alive:

1. **Stay calm.** The key response (as I've already said,

121

but it can't be repeated too often) in all hostage situations is to stay calm. Remember that you're dealing with people who've psyched themselves up for a caper. They're excited and scared, just as you would be if you were in their shoes. If they're veteran professionals, you're probably safe as long as you don't get too far out of line. But these days you're often dealing with amateurs. They're not cool. They say and do things without thinking or knowing why. It's dangerous—and the less you bug them, the better.

2.   Don't make wisecracks, question, lecture, or argue. In most cases, the only thing they've got on their minds is your money—and how to get out without a parade of cops following them. They could care less about your social theories, emotional responses, or personal hangups. All you'll do is slow things down—and increase your danger.

3.   When you're given an order, do it. Don't hang back or resist. These situations are always tense and hurried. Criminals don't have time to argue or reason with you. The least you can expect for your trouble is a cracked skull, and it goes downhill from there.

4.   Keep your head down. If you're taken along as hostage, keep your head down. Keep as much space between you and your nabbers as possible to minimize exposure to possible police sniper or SWAT fire.

5.   Learn the route. If you can see out of your abductor's vehicle, pay attention to the route taken—street names, landmarks, and businesses along the way. Your alertness may help the police if you escape or when you're released.

6.   Pay attention to details. There's usually a lot of time spent waiting around in these situations. Listen to what your abductors say to each other. Memorize their physical characteristics, dress, and speech, and listen for names (they'll probably use code names, but you never know when somebody will slip).

7. Don't try to escape. In the usual hostage situation, my advice is not to try to escape unless it's so easy that you can literally just walk away. In most cases, you're probably less likely to be hurt if you're patient. The record shows that most hostages are rescued or released unhurt.

8. Be very selective in the risks you take. If you should find yourself in an exceptionally dangerous situation, don't just panic and bolt. Bide your time. If you're in a car en route somewhere, look for an opportunity to make a break for it at a stoplight, while the car is stopped. Be alert for passing police; try to get their attention in some way; then make your move. It's better to do nothing than to take chances that are certain to result in your being hurt.

Airplane Hijacking

Airplane hijacking situations pose special problems. Every once in a while you read about some hero who subdues a hijacker 30,000 feet above Peoria. I never know which is the bigger fool, the hero or the hijacker. If a hijacker tells me he has a bomb and is going to blow up my airplane, I'm going to believe him. He may be romancing me, but at 30,000 feet I just don't feel like finding out. I'd rather take him where he wants to go, even if I do have to buy a box of El Ropo cigars to get my boarding pass back. That's why I'd advise the following behavior:

1. Keep your cool. Again, keeping your cool is of primary importance. Stay in your seat and let the crew deal with the problem, particularly in the case of inflight incidents.

2. Don't be a hero. Do what you can to discourage the people sitting around you from trying to be heros. If they fail, they are risking their own lives, your life, and as many as 200 or more other lives.

Political Hostages

Political hostage situations are probably the hardest to reason your way out of—and potentially the most life threatening—of all. Terrorists are not rational people; they are totally out of touch with the real world. Prevention is your best defense. Whenever possible, avoid traveling to countries where you'll be exposed to the possibility of terrorism.

If you do become a victim, follow the same guidelines you would in the life-threatening criminal hostage situation. Keep yourself open for any opportunity to escape. Politically inspired kidnappers are usually inexperienced criminals, but remember, amateur criminals can be the most dangerous.

Getting out of the trunk

One of the favorite tricks kidnappers and hostage-takers use to transport victims from place to place is to lock you in a car trunk. If you know what to do and can get your hands on something that will serve as a small screwdriver, you can open the trunk from the inside. Use a penknife blade or small screwdriver—there's usually one in the car's emergency tool kit also located in the trunk—to remove the small screws holding the plastic trim over the lock mechanism. Once the plastic trim is removed, the mechanism can be released from the inside with the same screwdriver or other implement. There may be slight variations from model to model and from make to make, but most newer cars have fairly similar trim and trunk latch mechanisms.

For practice, pull the plastic trim off your car's trunk lock, and play around with the lock latch until you've figured it out. Or have your car dealer or mechanic show you how. It won't cost much, and you never know when this kind of know-how might come in handy.

Another option for your own car is to have an inside trunk

release installed. This is a good safety precaution against children accidentally locking themselves in while playing around your car and also in the event that you are locked in by would-be kidnappers.

House Security
(See also Alarms, Home and Business; Apartment Security; Burglary; Doors; Windows.)

With the possible exception of our workplaces, most of us spend more of our lives at home than anywhere else. In the sections on robbery and burglary, we've already seen how high the risk is of having crime happen to us at home. These facts tell us just how important house security is to our personal protection. The rate of crime against houses, especially burglary, has more than tripled since 1960. If you live in a house, your chances of being ripped-off are getting better and better. Most of us can't afford the expensive protection of moats and alligators, but there are still a lot of things we can do to lower the score of the takers.

BUILDING SECURITY INTO YOUR NEW HOUSE

There are a number of advantages to designing security into your new home. First, you get better quality protection at little additional out-of-pocket expense, since the costs of hardware and installation are included in the price of the house. Second, security is better because it's built-in. Doors and windows can be selected and installed for their security value. Even the placement of window and door openings can be designed with the setting and other factors in mind to increase security.

Security Items to Consider

In recent years, design considerations in response to the rising tide of crime have become more and more important to architects. If you're planning to build, discuss your security requirements with your architect and builder. Ask them to include costs for security items such as the following:

- High-security doors and windows

- Keyless entry systems as an alternative to conventional locks on exterior doors

- Remote-controlled garage doors

- Full-perimeter alarm system

- Special security lighting, indoors and out, with lighting control center

- Electric eye or other warning system for your driveway to announce approaching vehicles

- Closed-circuit TV for 24-hour surveillance of grounds and entries

- Eye-level wide-angle viewers (or peepholes) for exterior doors

UPGRADING SECURITY IN OLDER HOUSES

Older houses present more of a challenge. They usually have ground-floor and basement windows to protect. Their doors and windows, as well as the locks, were installed when the only thing their designers worried about was the weather. Outdoor lighting is usually limited to porch lights and maybe one light outside the garage. Combined with the fact that they're very often screened from view by shrubs and trees, the vulnerability of older homes makes them a favorite target of intruders.

If you're trying to live in the world of the 1980s in a house designed for the world of the 1950s, my advice would be to do some things that will bring your crime prevention quotient up and your chances of being a victim down. Here's a checklist to guide you in upgrading the security of your house:

1. Outdoor lighting. The outdoor lighting for houses built before the 1970s usually does little more than prevent guests from falling off the porch into the rose beds. Get some cost estimates for outdoor lighting that will really light up your grounds and driveway.

2. Burglar alarm. Considering the small percentage of homes with burglar alarms (currently about 1%) you'd think installation of these things was a bear. It's not. Almost any one of the millions of do-it-yourselfers can buy and install an economical home alarm system that will give years of protection against the majority of break-ins.

Pick up an alarm to fit your budget this weekend and install it. It'll pay for itself 10 times over the first time it sends a neighborhood burglar packing. And who knows, it may even help discourage some kid from setting out on a life of crime.

3. Windows and doors. Take a trip to your local building materials store and talk to the boss about securing doors and windows. Explain that you want to harden your doors and windows against break-ins. Ask about the latest in preassembled doors and windows with frames that will fit the openings in your house (be sure to take measurements along when you go).

4. Locks. If the doors and windows in your home are reasonably strong and secure, check the dead-bolt locks on all exterior doors to see if they are in good working order. Make sure both the lock itself and the jamb plate are mounted on solid material by force-proof screws at least 2 inches long.

5. Landscaping. One economical way to protect your property from trespassers, vandals, and other intruders is to plant thorny shrubs and hedges in strategic places. Plant them around walks and windows, but keep them trimmed. Thick, chest-high hedges make good fences against burglars.

Ask someone in a nursery or garden center which throny shrubs are best suited for your soil and climate conditions. When putting in new plants, set them at an angle of 45° to one another. As they grow, wire the branches where they cross. This technique, known to the *green thumbs* as "splashing," will provide you with an impenetrable barrier to would-be intruders.

6. Hinges. If local ordinances dictate that doors should open outward, requiring hinges and hingepins to be on the outside of the door, be sure to peen (flatten) down the end of the pins to prevent easy removal of doors from their frames.

7. Sliding glass doors. Secure these as described in the sections "Doors" and "Locks."

8. Gadgets. The same range of electronic gadgets available to the new home builder is available to you. They're at your local electronics store for do-it-yourself installation. These include lighting centers to control all the lights in the house, intercom systems to keep track of what's going on in the rest of the house (these can be audio, video, or both) and individual door and window alarms.

9. Dogs. No home security system is complete without a dog. They make good pets for the kids and, if reasonably doglike, will bark their heads off if any stranger sets foot on their property. If you live in a house, it's generally easier for you to care for a dog than if you live in an apartment. And don't think you have to have a Doberman or German shepherd; a mutt will do just as well and will probably live longer. So take a trip to your local Humane Society and invest in some economical house security that's also support for a worthy cause.

J

Jogging
(See also Mugging; Personal Safety; Self-Defense.)

What's a nice, safe subject like jogging doing in a book like this? Well, as you know, millions of Americans are into the sport. And as their numbers grow, so do the chances of runners being assaulted, robbed, raped, and seriously injured. Although the popularity of jogging is national, it's biggest in metro areas. And, unfortunately, inner-city streets, where most violent crime occurs, are frequently the only routes available to serious midtown joggers.

Because of the age groups involved—adults of both sexes between 18 and 44 who see themselves as the "white hats" —there is a tendency to dismiss the risk as easily manageable or nonexistent. That's just plain foolish. The majority of violent criminals, inside and outside prisons, are in that age group too. So we're just kidding ourselves if we don't recognize the problem. It's dangerous on America's streets, and that's where we jog. If we want to protect ourselves and each other, we have to face that danger and deal with it. You might say there's a time to jog and a time to run like hell!

11 RULES OF THE ROAD

1. **Plan your runs carefully.** Make it a habit to run in groups of two or more instead of running alone. And stay out of known high-crime areas or places where muggers are known to congregate, such as public parks.

2. **Dress down.** Don't wear expensive jewelry or watches that are likely to attract muggers.

3. **Carry the minimum.** Carry identification, but not your wallet or keycase. There are robbers who will grab a jogger then hold him or her while one of their associates goes to the jogger's home and cleans it out.

4. **Run like hell.** If you do run into trouble, don't attempt to slug your way out of it. Just get the hell out of there. I mean, wail! Even if you're Black Belt, I wouldn't recommend bare hands against creeps with guns, knives, brass knuckles, and steel pipes.

5. **If trapped, stay cool.** Don't lecture, threaten, or act superior. If it's a gold chain, a bracelet, ring, or wristwatch he's after, give it to him and get him away from you. Don't prolong the encounter. Your life is worth more than baubles. Save your eloquence for the police; tell them everything you can remember about your mugger; and let them handle it.

6. **Carry alarms.** Carry an electronic noise alarm or fog horn to scare off would-be muggers.

7. **Always run facing oncoming traffic.** This is for your physical safety as well as to spot cars with suspicious-looking characters.

8. **Carry emergency money.** Make sure you have change for the telephone in case you're lucky enough to be near one if you need it.

9. **Run during daylight hours only.**

10. **Be alert.** Listen and watch for any unusual or suspi-

cious activity around you when you run. Report anything strange to the police.

11. Watch out for strangers. Be wary of strange joggers who suddenly show up during your runs. Just because they wear jogging suits doesn't mean they're automatically okay. Jogging is being used more and more frequently as a cover for casing and committing burglaries and robberies, particularly in suburban residential areas. Report any oddball joggers you meet to the local police for security checks. If they're legitimate, they won't mind showing IDs.

K

Kidnapping
(See also Hostage Situations; Personal Safety.)

A lot of people think kidnapping is a problem strictly limited to rich kids and their parents. But it's actually much more widespread in the U.S., with 205 kids reported missing every hour. That's 1.8 million kids a year. Many of these kids are runaways. Others are snatched by one or the other of their divorced or separated parents. And though many are later found or turn up safe, about 50,000 disappear forever, every year.

It's true that if you wield wealth and power, your children are in greater danger of being kidnapped for ransom. But there are many reasons for kidnapping, and ransom is only one. Anyone can lose a child if he or she is careless. In fact, by far the largest number of the 50,000 annual disappearances are from families that are neither rich nor powerful. The kidnapping of a child is almost always due to parental negligence, inattention, and carelessness.

Babies are the most easily protected. There's really no legitimate reason why an infant should be out of sight of its parent or nurse at any time. Children are another story. It's harder to protect them if they're to have anything like a normal childhood. And since you can't be with your children 24 hours a day, it's important that you educate them early to the dangers of kidnapping. When I talk to my kids about kidnapping, I try to enlist their cooperation by carefully explaining the problem to them and being very clear about the things they have to do to avoid it.

EDUCATE YOUR CHILDREN

There are several pointers for you to follow in order to minimize the dangers of kidnapping:

1. Be alert to the danger. If your political position or wealth makes your children a likely target, explain to them the reasons why they should be especially cautious. Don't leave it to some creep to surprise them. Make sure they understand all the reasons for being super careful.

2. Be honest. Tell the kids the good and the bad of the society around them. Make sure they see the need to discriminate and not simply to assume that everyone is okay.

3. Stick together. Explain the need for people, particularly small ones, to stick together and take care of each other. Impress on them the safety in numbers principle and the increasing danger of going alone.

4. Steer clear of strangers. Ask your children to watch out for adults who may try to lure them into cars or to get them to accompany them someplace. Tell them to report any of these attempts to you immediately.

5. Vary schedules. If your kids attend school, suggest that they vary their schedules to that they don't leave the house or return from school at the same time every day. If possible, have a responsible adult accompany them to and from school.

6. Make up a secret code. Use a list of code words to communicate with one another without tipping off kidnappers or others to what's being said between you.

7. Keep in touch. Make up a game in which they have to keep you posted as to their movements. One of the rules of the game would be that they have to call you on the telephone when they arrive and again before they leave to return home. If possible, set time schedules for the calls.

8. Check the bus. If they ride with neighborhood kids on a school bus, make sure one parent is always there to see them off and to meet them when they return. Rotate this responsibility among the various parents in the neighborhood.

IF KIDNAPPERS STRIKE

Here are some tips just in case a kidnapper does get into your life:

1. Don't panic. Try to reassure yourself with the knowledge that *about 98% of all kids reported missing are found*. Violence in kidnapping, although heavily played up by the media when it does happen, is, in fact, rare. In other words, most kids reported missing eventually turn up safe. This fact should in no way soften your determination to prevent it from happening. That 2% figure is still far too high a price to pay for carelessness.

2. Call the police. Do this the minute you begin to suspect that your child may have been kidnapped. Give them your name, address, and phone number and the child's name, age, and description. Don't feel embarrassed if your child later turns up safe. It's better to jump the gun than to give a kidnapper time to get his act together.

Many police departments have special missing children details. They have experienced detectives who can probably tell you very quickly whether you've actually got a kidnapping on your hands, or if it's more likely that your ex has the child, or that your child ran away, or is lost.

3. Talk to your child. If you're hit for ransom by someone who claims to have your child, demand that they put the child on the telephone as proof. Then find out what their demands are. When they've made their pitch and you have their instructions, call the police and give them full details of the conversation, such as instructions, background noises, and peculiarities in the voice or pronunciation of the caller.

If the caller warns you not to call the police, ignore him. You're dealing with a criminal, and you have no reason to trust his word on anything. If indeed he has your child, the chances are better than even the police will get your child back unharmed. Very few ransom schemes work; almost all result in capture of the kidnapper when police are able to act in time.

Kids
(See also Drugs in the Second Degree; Kidnapping; Vandalism.)

Forty years ago, if a tough street kid didn't like your short pants, he might throw a rock at you or punch you in the mouth. If you punched him back, you probably had a fistfight on your hands. Today, if a kid gets into a fight with some punk, he's liable to be stabbed or shot. The Marquess of Queensbury is out the window.

Ah, you say, that only happens to teenagers and young guys looking for trouble. I've got news for you: it's happening to kids—little kids.

In 1964 about 328,000 kids under 18 were arrested for

serious crimes: criminal homicide, manslaughter, rape, robbery, aggravated assault, burglary, larceny/theft, weapons possession, and narcotics violations. By 1982, this number had jumped to 719,464. And what's worse, we've recently opened up a new category in our crime reports—*arrests under age 15*. During 1982, 550,901 children under age 15 were arrested for serious crimes, including 5,456 for carrying guns and 11,405 for narcotics violations. And these are just the kids that got caught! They probably represent well under half the total number of kids who are getting into serious crimes.

Forty years ago, I would have made damned sure my kid punched anyone back who picked a fight with him. Today, I'd tell him to walk away and avoid street fights or arguments of any kind. That's his best defense, since I don't feel justified in letting him walk around carrying a knife or a gun for his protection, and it's becoming obvious that politicians and judges can't keep kids who do off the streets anymore.

STREET SMARTS FOR THE SMALL FRY

Here are a few tips on how your kids can avoid street crime:

1. **Stay away from strange kids.** Tell your kids to do whatever's necessary to stay away from strange kids, whether it means changing sides of the street or walking around the block to avoid them. Rather than argue or fight, tell them to ignore insults made by other kids trying to provoke them. Let them know that it's better to be called "chicken" than the alternative: getting shot or stabbed.

2. **Don't let your little kids hang out.** Keep them off the street and busy with activities, chores, schoolwork, play at home, and hobbies.

3. **Talk about the dangers of using drugs.** Remind

139

them that fooling around with drugs is dumb, a sucker's game. Ask them to tell you if friends or others offer them drugs, including alcohol.

4. Stay out of cars. Tell them to stay out of cars driven by kids they know are too young to have driver's licenses, as well as by strangers. Make sure they know that they should walk on the side of the street facing traffic and to be on the lookout for anyone who might stop a car and try to force them into it.

5. Tell them to make a lot of noise. Tell them to steer clear of strangers, to run from anyone who tries to grab them, and if anyone comes after them, to scream loud and long. Most of the freaks who dig kids are not interested in having an audience. They'll pass if your kid makes enough noise to attract the attention of other adults and stays out of reach.

6. Tell you what happens. Make certain they understand that they have to tell you immediately if *any* adult molests them sexually or suggests anything "funny" to them.

7. Teach them how to use a pay phone. Do this as soon as they're old enough, and make sure they always have emergency money. They should memorize the emergency police number to call if they ever need help and know how to report their location using the names of streets at the nearest intersection, an address on a nearby building, the company name of a nearby service station, and so on.

GETTING YOUR TEEN'S HEAD STRAIGHT

Because of peer pressure and adolescent insecurity, teenagers are more likely to become involved in crime than any other age group. Over a quarter of the people arrested for crimes in 1982 were between the ages of 13 and 19, with over a half-million arrests in each age group from 17 thru 19. And more 19-year-olds were arrested than any other age group in the population.

Drugs, including alcohol, are the No. 1 threat to your teen's future. Kids are under enormous pressure to use marijuana and cocaine, both of which are commonly found around bars and hangouts and even in schools, parks, and playgrounds. Unless they have parents who have warned them about the dangers of drug abuse, kids may see no reason not to take drugs when they hear about all the disc jockies, rock stars, Hollywood celebrities, sports personalities, and jet setters who are into drugs and whom their friends and mentors envy and admire.

It's up to you as parents to counter this bad-news trend. Tell your kid the truth about drugs. Don't lose him or her by default. If you need help, consult a drug clinic (see ''Drugs'').

Tips to Avoid Street Crime

When it comes to street crime, I give teens a lot of the same advice I give the younger kids.

1. **Back off from street fights.** Too many kids are carrying knives and guns.

2. **Don't take dares.** Don't be intimidated into doing dumb things.

3. **Let them take it.** If some kid comes up and grabs your bicycle, stereo, or money, don't argue or fight; let him have it. But remember what he looks like, how he's dressed, and any unusual things about him that will help the police identify him and pick him up.

4. **Don't show off.** Even if you're with your friends, don't try to show them what a tiger you are. Avoid confrontations with strange kids in school hallways, playgrounds, and neighborhood streets. They may be Eagle Scouts out earning merit badges; but chances are they're not. Load the Volkswagen and boogie on out. Don't hang around to ask questions.

5. **Don't be too "in."** If you want to be *really* cool,

avoid things that are too "in." Rock concerts are a good example. Apart from the fact that you get ripped off at the box office, there's also the problem that these things attract just about every screwup in town. They're usually promotions carefully designed to exploit kids, and they're almost always overrated and overpriced. Often, drugs are sold openly, a lot of beer flows, and kids get into fights. More and more kids are getting hurt, becoming victims of assault, or getting raped at these events. Cops go to them now, and kids out for a good time wind up arrested.

6. **Avoid booze.** Don't get into the habit of having a few beers with your friends just because they expect you to. The young kid who has to have a few beers today is the older kid who has to have a few belts tomorrow. You've got to be sharp to be a survivor in today's world, and the kids who are getting drunk are going nowhere at the speed of light.

Larceny/Theft
(See also Airport Crime; Car Stripping; Car Theft; Credit Cards; Pickpockets; Purse Snatching; Retail Store Crime; Social Security Theft; Vacation Travel Tips.)

Your exposure to larceny depends on how alert you are to the risks. More than any of the other crimes I talk about, this one owes its popularity and success to our failure to recognize it and take it seriously. Thieves profit by exploiting that shortsightedness.

All of us have sinned or been sinned against through larceny. When we were little kids we "lost" toys. During our teens, our bicycles were stolen. And as adults, our cars are stripped, businesses ripped off, tools, equipment, or supplies stolen, cash embezzled, and possessions taken from our homes.

And on the flip side of the coin, we take things from hotels, motels, and restaurants as souvenirs—and then pay for them in higher rates and prices. Supermarkets and department stores lose millions each year to theft; we pay it back in next year's food and clothing bills. The construction industry loses millions in stolen building materials and equipment; we pay for them in rents, mortgages, and insurance rates. It's a vicious circle, and we all end up paying for it.

The estimated cash value of all this stealing is about $2.5 billion a year, and that's just for openers. Officials believe actual losses—which take into account tax revenue losses, reimbursements, and rate and price increases—are probably many times higher.

What we're looking at is a tremendous drain on an economy that's already leaking from every seam. Look at it this way: Honesty is still the best policy—not just because it's right, but because it costs less.

CHECKLIST: HOW VULNERABLE ARE YOU TO THEFT?

Do you

- Ever leave your kids' toys, bicycles, baby carriages, bassinets, etc., in the hallway outside your highrise apartment?

- Fail to lock your bicycle at school, at work, or out on the street?

- Neglect to lock up your vehicle, or protect wheels, and other vulnerable areas?

- Believe that alarms and security devices are just too much trouble?

- Feel that strict money control and other accounting procedures to reduce theft interfere with the way you serve your customers and your relationships with your employees?

- Leave tools, equipment, vehicles, or materials and supplies outside your home or business?

- Let mail collect in your unlocked mail box?

- Leave your garage door open during the day?

- Own an unprotected vacation home?

- Fail to count your change or review charge account receipts?

- Pay restaurant checks and bar bills without verifying charges?

- Leave valuables, watches, jewelry, cameras, binoculars, portable radios, and TVs carelessly exposed on desks or tables in your office, home, or apartment?

If your answer is yes to any of these or similar questions, you are about to have something stolen. *Preventing it is up to you.*

SOME IDEAS ON HOW TO AVOID THEFT

At Home

1. **Don't tempt visitors.** Don't leave small, expensive items around your house or apartment where casual guests or visitors, parading small fry or others with sticky fingers might succumb to their charms and relieve you of them.

2. **Keep valuables hidden.** Keep wallets, purses, and cash in drawers when not in your hands or on you.

3. **Store toys inside.** Teach your kids to bring skate boards, tricycles, bicycles and other toys inside when they've finished playing with them.

4. **Put it away.** Set a good example by putting your own tools and materials away.

5. **Check your mail.** Check the mailbox every day for letters and packages.

6. **Lock it up.** Keep your garage door closed whenever possible. Don't leave it standing open during the day.

7. Lock your car whenever you leave it. And don't leave valuables in it overnight, particularly if it's parked on the street or in a public garage.

Away from Home

1. Lock up your car. Secure your car with locks on the wheels and covers, a locking gas cap if the filler door didn't come with a lock, and hood locks. (See also "Alarms, Auto.")

2. Hide your stuff. Don't leave shopping purchases, liquor, or other merchandise in your unlocked car in full view of passersby. If the car has a trunk, keep your loot locked in it; if it doesn't, keep an old blanket handy for covering the stuff on the seat.

3. Use the glove compartment. Stash sunglasses, binoculars, tape cassettes, cameras, and other small valuables out of sight in the glove compartment. Lock it if you're leaving the car for a few minutes, or take them with you. Don't leave them in the car overnight.

4. Protect your business. If you own a business, take a survey to determine your security needs. Then purchase the equipment necessary to lock up and secure all valuable property such as typewriters, copiers, dictating and transcribing machines, trucks and vehicles, fuels, valuable tools, and machines. If employee theft or pilferage is becoming a problem, talk to a corporate security systems consultant (look through the listings in the *Yellow Pages* under Security Systems Consultants).

5. Watch your cash. If you're in a business where "till-tapping" is getting out of hand, you can discourage the "tappers" by instituting strict cash management routines to limit cash in drawers and by tightening daily accounting procedures.

6. Travel light. This is true whether you travel for

146

Locks

business or for pleasure. Take as few valuables as possible. Keep travel cases, attaché cases, cameras and binoculars over your shoulder or under your arm at all times. Don't put them down where you might forget them or fail to miss them if they're taken.

7. Don't flaunt it. Whether you're traveling to a business appointment or to a vacation resort, resist the impulse to flaunt your affluence. Don't deck yourself out in the best things your money can buy if you're interested in keeping them for that special occasion. Old luggage, battered portable radios, scuffed briefcases, grubby trenchcoats, old golf clubs and well-worn camping equipment will attract fewer takers than shiny, new, and expensive stuff.

Locks
(See also Alarms, Home and Business; Apartment Security; Burglary; Doors; House Security; Personal Safety; Safes; Windows.)

**Question: When is your lock not a lock?
Answer: When it's unlocked.**

In other words, don't install a bunch of expensive locks and then forget about them. You're liable to lose a lot more than just your hard-earned bucks. Buy locks, install them right, and keep them locked.

Some people I meet throw up their hands when I start talking about locks, as though they are just a waste of time. I tell them not to sell locks short, particularly good dead-bolt locks with high-security cylinders, which I'll tell you more about in a minute.

147

About 20% of all residential burglaries in 1982 were walk-ins, or illegal entries. That represents a loss of over $650 million just because a lot of people didn't lock their doors!

Starting right now, we can reduce the incidence of burglary—the nation's second largest crime category—by 20% simply by always locking our doors. And that 20% reduction would bring the nation's overall crime rate down by six percentage points!

DOOR LOCKS: BAD, GOOD, BETTER AND BEST

Bad

If you've recently moved into a new house or apartment that has key-in-knob door locks (Figure 11a), you won't be happy to know they're basically useless for home security. The complete tool kit of the jet-age burglar consists of a thin screwdriver: That's all he needs to get through the average residential door in seconds. If you have key-in-knob locks on your doors, change them or install one of the backup or auxiliary locks I discuss below.

Good

Install a rim dead bolt, which can't be easily shimmed, loided, or jimmied (Figure 11b). Rim locks are easy to install on existing doors and jambs, and they offer greatly improved security when used with conventional key-in-knob locks. Avoid spring latch rim locks, the so-called snap locks (Figure 11c); they are easily shimmed or loided.

The disadvantage of any rim lock is that unless the rim jamb plate is reinforced with heavy-duty fasteners, the jamb plate is easily broken loose by a hard kick or butt from a determined, hard-headed intruder.

Better

You've heard me say it before: Don't put a $50 lock on a $10 door. If you're upgrading your doors from hollow to

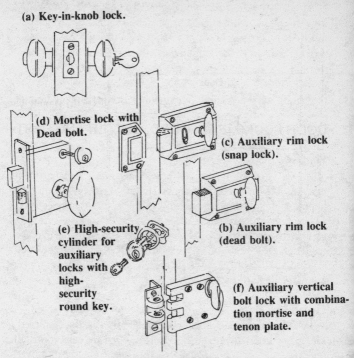

(a) Key-in-knob lock.

(d) Mortise lock with Dead bolt.

(c) Auxiliary rim lock (snap lock).

(e) High-security cylinder for auxiliary locks with high-security round key.

(b) Auxiliary rim lock (dead bolt).

(f) Auxiliary vertical bolt lock with combination mortise and tenon plate.

Figure 11. LOCKS.

solid core, now's the time to chip out a place for a good quality, dead-bolt mortise lock (Figure 11d). But don't bother unless you have solid doors. A strong lock on a hollow core door is like a bullet-proof G-string. It may look great, but it doesn't offer much protection.

Unlike rim locks, which are attached to the outside surfaces of both the door and the jamb, mortise locks must be hand-fitted into the door itself. A wood chisel is used to carve a precise hole the exact size and shape of the lock. The lock

LOCKS (Cont.)

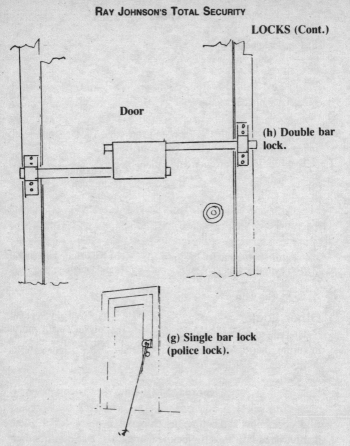

Door

(h) Double bar lock.

(g) Single bar lock (police lock).

is then fitted into the door and secured flush with the edge of the door with screws. In effect, it becomes part of the door. Top quality mortise locks, now frequently combined with high-security lock cylinders or keyless locking systems, are among the oldest and best of the door locks (see Figure 11e). Castle keeps of yore had beamed doors with great iron dead-bolt mortise locks.

Best

There are two categories of top security, auxiliary bolt-on locks:

1.   The vertical bolt lock is actually a combination mortise and tenon plate lock in which twin bolts fit through holes in two tenons projecting from a jamb plate that is attached by long screws to both the face and the edge of the door jamb and joist (see Figure 11f).

2.   Bar locks provide maximum security for the embattled apartment dweller. These locks come in a wide variety of designs and models, from the steel bar police lock in Figure 11g, with one end fitted into a metal groove in the floor and the other end held in place by a locking mechanism on the door, which can be activated or released (in some models) by a key from the outside; to double bar locks such as the model in Figure 11h. For the truly paranoid, bar locks are available with as many as six bars.

Other Locks

There are a number of other bolt-on locks available, including battery-powered, alarm chain locks, and various combinations of rim, mortise, and tenon locks.

I don't recommend the so-called safety chains. Most of them are ineffective, come with attaching hardware meant for model airplanes, and have chains that can be easily severed with bolt cutters. They're dangerous because they give the illusion of safety until the moment some turkey gives the door a quick shove and walks on in.

KEYLESS ENTRY SYSTEMS

The keyless entry systems, which are state-of-the-art systems, are expensive. The door locks themselves are usually of the mortise type, which is also available with the much less costly high-security key lock.

Since only a very small percentage of burglars know how to pick locks, the additional security of a keyless lock system is probably not worth the additional cost. Then there's the further disadvantage of the need for regular maintenance (frequent checks, battery replacements, etc.).

Banks, large companies, and a few exclusive condominiums and hotels are into card key entry systems using magnetic card keys sandwiched in plastic. Some of these systems are pretty effective, whereas others can be bypassed easily. They're all expensive and impractical for conventional residential purposes.

TIPS ON BUYING AND INSTALLING LOCKS

1. **Survey of your house or apartment.** Determine which doors have adequate locks and which require upgrading with auxiliary locks. Then take an inventory of your belongings. Which things are the most valuable? The most likely to attract a thief? Are they adequately protected by locks?

2. **Check for proper protection.** Are all outside entries, including doors leading to attached or lower-level garages, well protected with jimmy- and pick-proof locks? (These doors are the favored private entrances of burglars.) All of these doors should be protected with good-quality dead-bolt locks.

3. **Safeguard your garage.** If your lower-level garage door opens out, as is often the case with these doors, chances are that the hinges are on the outside too. This means you'll have to peen down the hinge pins or, better still, install dowels to prevent door removal in case the burglar does manage to remove the hinge pins. (See Figure 8, in section on "Doors.")

4. **Inspect locks on all windows.** Replace or repair any that don't work.

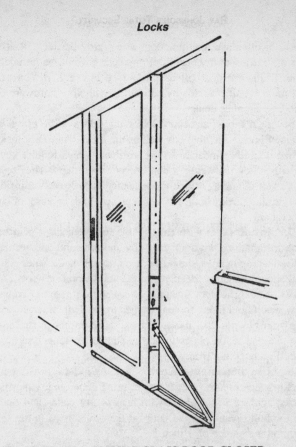

Figure 12.   SLIDING GLASS DOOR CLOSER.

Also functions as lock and prevents door from being lifted off its tracks.

5.   **Pay special attention to locks on sliding glass doors.** If anyone decides he wants your cherry wood breakfront, this is the door he'll take it through. One model of sliding glass door has a closer system that prevents the kids from leaving

153

the door open and also functions as a sturdy bar lock. It also has the extra security benefit of preventing would-be intruders from lifting the door off its tracks (see Figure 12). You can get more specific advice by consulting with your hardware or building supplies dealer.

6. Don't put auxiliary locks on doors with glass in them. Replace the doors as well as the locks. Auxiliary locks called double cylinder locks are available which require keys on either side. However, I don't recommend them. They tend to be self-defeating, since the tendency is to leave the inside key in the lock, both for convenience and in case of an emergency.

7. Install them with care. Make sure your do-it-yourself lock installations are done carefully and are solid and secure. All attaching screws should be a minimum of 2 inches long and should penetrate deeply into the solid wood of doors and joists around the door opening. Don't settle for short screws; they won't hold if real force is applied to the door. If necessary, use your electric drill motor to make holes larger in strike or mortise and tenon plates to accommodate the longer screws. (See Figure 9 in "Doors.")

8. Use maximum security lock cylinders that resist picking (see Figure 11e). If you're using good quality mortise locks, replace strike plates with heavy duty plates and open up the bolt opening in the jamb so the bolt can go in the full minimum distance of 1 inch.

9. Take care of your keys. Make it a habit to carry only those keys you actually use, and never hang keys on walls or leave them lying around on tables or cabinets. Decide on a secure place to store keys and make sure everyone who uses them understands the importance of keeping them there.

Remember, the most important feature of any lock is your determination to use it, and your own and your family's success in developing good lock-and-key habits. There is no more effective way of keeping you, your family, and your property safe than installing good locks *and keeping them locked*!

M

Mugging
(See also Airport Crime; Apartment Security; Armed Robbery; Automatic Teller Machines; Drugs in the First Degree; Elderly, Common Crimes Against the; Freeway Crime Smarts; Jogging; Murder; Personal Safety; Push-Ins; Rape; Self-Defense.)

Uniform Crime Reports defines muggers as small-time robbers who use force or the threat of force to take things from people. But in fact, muggers can also be rapists, murderers, assailants, and a whole array of dangerous criminals. The majority of muggers are young, inexperienced, ignorant kids, looking for easy marks. They're particularly dangerous to young people, lone women, the elderly, and the infirm.

The first question you need to answer is this: How much of a threat does mugging really represent in your area? In general, big towns represent big risk, little towns, little risk, but there are exceptions. If you live in Manhattan, Kansas, and plan to stay off of Manhattan Island, New York, you may want to skip this section. But if you live in a decidedly populated area where mugging is a potential problem, read on.

STRATEGIES FOR HARDENING THE TARGET

Most of us know when we're going into an area where there's a risk of being mugged. When we leave the security of our apartments or suburban homes for the streets of the South Bronx, we just *know* we're going to be dodging knives and bullets. Instead of sweating a lot and hoping it won't happen to us, this is the time to do some realistic thinking and prevention homework. There are some general guidelines you should always follow:

1. Don't go alone. The old adage about safety in numbers still goes. Don't walk through dangerous neighborhoods alone. If you must be in high-risk areas, go in groups of two or preferably three or more, particularly if you're a kid, a young teen, elderly, or a woman.

2. Take the safest route to your destination. Decide on your strategy in advance and stick to it. Unless there is absolutely no other way to reach your destination, don't travel through areas of the city that you know are dangerous; go around them.

3. Don't dress up in your Sunday best. Muggers look for the well-dressed, easy mark. If you're in town shopping or seeing your dentist, wear casual, inexpensive-looking clothes.

4. Keep your antennas out. Be alert. If you see guys that even *look* like muggers, get it into gear. If you can, run like hell. If not, walk around the block quickly, or duck into an open shop. The point is, keep a lot of real estate between you and them. Don't just grit your teeth and walk into their arms.

5. Arm yourself with a well-tested personal protection alarm. If it's one of the electronic types, check the batteries regularly. If it's the aerosol type, blip the button to make sure it has a good blast to it and plenty of air in reserve. Carry one of these surrogate shriekers in your coat pocket or handbag, ready for instant use. Most muggers won't want to hang around when you've got one of these things going bonkers in your hand.

6. Travel in the morning. Schedule appointments and errands during morning hours. Avoid dangerous areas during the late afternoon or after dark.

If you can't avoid going into high risk areas alone and at odd hours, here are some don'ts.

1. Don't flaunt wealth. Don't carry large amounts of cash; don't wear valuable jewelry, chains, or watches; and don't carry touristy items likely to attract attention such as expensive cameras and stereo radios.

2. Don't carry a purse. Keep your ID in a pocket and carry larger necessities in a brown paper or shopping bag. If you must carry a purse, carry *it* in a paper bag.

3. Don't fight back. If muggers do get to you, in spite of all your precautions, don't try to fight them off alone. Don't bug them and don't try to bargain with them. Every once in a while you read about some cat who's up for a black belt, or maybe an 80-year-old grandmother, who beat the bippies out of a would-be mugger and held him down till the cops came. It makes better copy than the other 50 victims who tried their luck and wound up with extra air holes for their trouble.

Sure, there are some patty cakes running around out there, only pretending to be bad guys. But it would be just your luck not to pick one of them. So, unless you're Joe Frazier or Larry Holmes—or even if you are—my advice is to give the punks what they want and get them away from you *muy pronto*. Most of them are young and nervous. They'll use that knife or gun without thinking.

WHAT IF YOU'RE MUGGED?

In the unhappy event that you get mugged, remember that the time to go after the muggers is when the cops arrive. It will help if you can give them a good description. Often muggers are well known in a particular neighborhood or have been previously picked up in the same area by police. So your careful description and identification might result in a quick arrest.

Whatever you do, don't just walk away and keep the experience to yourself. Remember everything you can about your mugger, and make a full report to the police. Give them all the help they need to get those guys off the streets and put them away.

Murder
(See also Guns; Personal Safety.)

There is no crime we dread more than murder. Yet of all the bad things that can happen to us, murder is the least likely. Murder represents less than two-tenths of 1% of known offenses. Why then are we so preoccupied with it?

Partly because the media spends so much time and space making sure we don't forget it. And partly because, of all the things that could happen to us, murder is the most permanent.

Here are some statistics to think about: A little over 15% of all murders are between strangers. Some 30% occur between people who may have known each other. But the majority, at least 55%, are between people either related to or known to one another. Significantly, 42% of all murders result from arguments; and drinking is an important factor in many arguments.

What these numbers tell you is that considering the slight chances of being murdered in the first place, your chances of being murdered by a stranger are very low indeed. If it did happen, the culprit would probably be a relative, a lover, a friend, someone you know through your work, or someone you socialize or drink with, and the crime would very likely occur as a result of, or during, an argument or fight.

SIGNS OF MURDER

Murder is a social crime. And though laws can keep guns out of some people's hands, only people can prevent murder.

Like much of the crime discussed in this book, the majority of murders can be prevented; though, of course, there are always a few that can't. Some crime will always be with us—until we design the perfect human being and the perfect world. But it is within our power (if we use our common sense) to prevent a lot of crime from occurring.

I never killed anyone myself, but I spent 25 years of my life with guys who had and who spent a lot of time talking about it. Everything I heard makes me think that most people who become victims of murder have some warning well in advance. That seems logical since we know that most killers are known to their victims. With this as background, I've put together a few signs of what you can look for.

Between Husbands and wives

Almost 10% of the murders reported are between husbands and wives. High on the list of causes are sexual problems, jealousy, competition, and arguments over money. If you and your spouse are having violent fights over these or other matters, back off and take a hard look at what you're both doing. Get outside counseling if you need it, but don't put off dealing with serious arguments and problems in the hope they'll go away. They could take one of you with them.

Between Friends and Acquaintances

If you're having trouble with friends or business acquaintances, regardless of how petty or unimportnat it may seem to you, my advice is to give the problem some thought. Then get together with the person and try to resolve it to your mutual satisfaction. A festering grudge or a sense of injury can sometimes lead to serious, perhaps life-threatening, complications. If you can't be reasonable with each other, you'd better look to your options.

People get killed over the most trivial things—from running the lawn mower too early on Sunday morning to arguments over the point spread on a $5 Superbowl bet. Don't let yourself get sucked into emotional battles. Think before you say things, and try to meet the other guy halfway. If common sense can't resolve the situation, get help. Don't wait for the turkey to come back with his gun.

4 COMMONSENSE WAYS TO REDUCE YOUR CHANCES OF BEING MURDERED

1.   **Beware of strangers.** Although murder between strangers is not common, it has been on the increase recently. So, be alert to any stranger, in the street or elsewhere, who tries to approach you for no apparent reason. And, as I've said many times, don't open your door to strangers.

2.   **Be courteous and discreet.** Don't go out of your way to be nasty to or to antagonize other people. Be considerate of others, whether you're behind the wheel of your car or in a social or business situation. In short, don't go around making enemies just for kicks. The odds against murder may be in your favor now, but you *can* turn them around in a hurry.

3.   **Don't put up a fight.** If you're confronted by a robber, burglar, or thief, give him what he wants and get rid of him. Don't fight, argue, or lecture. Keep it short, let him go, and don't follow him.

4.   **Be aware.** Above all, don't become complacent and convince yourself that you're so cool it couldn't happen to you. If you buy that one, you've got one foot in the box already. None of us is bullet proof. Murder may be rare in comparison to other crimes, but it still happens—sometimes right down the block.

P

Peeping Toms
(See also Personal Safety; Rape; Telephone Security; Windows.)

Peeping is a crime in some communities. Those communities also have laws against similar crimes such as trespassing, loitering, and violations of privacy. The peeping tom we're talking about here is basically the guy who sneaks around in jogging shoes peering into other people's bedrooms at night.

A lot of people start chuckling the minute anyone mentions peeping toms. They don't think of it as a crime; it's a kinky but harmless activity. Okay, when you compare it to all the other stuff that's going on, it may not seem like much. The problem isn't with the peeping itself, but with what it represents.

You've got some character watching you do whatever you do, every night. Why? What's his problem? Doesn't he have anything else to do, like fix the faucet or watch TV? Is he some kind of nut, and if so, is he really harmless?

WHO PEEPS? AND HOW TO AVOID IT

The morals squad of your local police will tell you peepers come in all sizes and colors, from college professors to tiny tots. A lot of peeping is actually done by curious or frustrated adolescents. These are usually the timid kids who haven't yet figured out how to make it happen in the back seat of a car.

But the other large category, the kinky adults, tend to be more dangerous. The long-range peeper who spends his nights in a dark, high-rise apartment looking through his telescope is probably nothing more than an emotionally immature guy who's bored with his wife or girlfriend. The real danger comes when you attract someone who wants to do more than peep and begins to make obscene phone calls and even unsolicited personal appearances.

There are a number of ways to avoid becoming the target of a peeping tom:

1. Cover your windows. Your best insurance against peeping is to have good blinds or shades at every window and to keep them drawn at night or whenever necessary. Make sure they're long enough to completely cover the window, and don't leave tantalizing little cracks at the bottom edges.

2. Use pressure mats. If you think peepers are hanging around your windows at night, pick up a couple of pressure mats to add to your home burglar alarm system. These are accessory alarm switches that are activated when stepped on and can be used out of doors. You can bury them in flower beds, gravel, redwood chip edging, and so on. Hooked into your noise and light alarm system, they'll light up your peeper's life the next time he bellies up to your bedroom window. Chances are, he'll look for another window to peep into.

3. Install outdoor floodlights. Aim them at areas with ground-floor windows. Then equip light switches with multiple cycle timers that turn lights on and off at random. Peepers, like other nocturnal creatures, hate the light.

4. Trim your greenery. If you have large trees on your

property, prune the lower branches up to a minimum height of 12 feet from the ground flush with the trunk to guard against arboreal peepers.

5. Get a dog. It doesn't have to be an attack dog. Any little mutt with a yappy bark will more than earn his keep by alerting you to the approach of strangers and sending them packing. Plus you and your kids will have a friend for life.

In most cases, you'll discourage neighborhood peeping toms with good blinds and outdoor lighting. But if the peeper persists, appears to be getting bolder, or attempts to make contact in any way, here are further steps you should take:

1. Notify the police. Don't feel embarrassed discussing it with them. Call the police even if you don't feel threatened. They'll be the first to tell you peeping isn't a joke.

2. Listen to the cops. Follow their advice and instructions.

3. Don't be a hero or heroine. Never charge out into the night to scare the hell out of the creep. You never know what you'll confront out there.

4. Avoid contact. As with other criminals, avoid contact with any peeping toms. Don't respond to telephone calls or notes, whether obscene or not, from peepers; discuss them with the police. And don't agree to any meetings or "dates" unless you're accompanied by the police or other security officer.

5. Be cautious and suspicious. Treat any peeping tom incident with the same caution and suspicion you would any other attempted crime. Don't just laugh it off. People are either straight or they're not. If he's into peeping, he's into a lot more than any straight person ought to be. He needs help. Let the judge get it for him.

RAY JOHNSON'S TOTAL SECURITY

Personal Safety
(See also Armed Robbery; Automatic Teller Machines; Burglary; Computer Dating Services; Freeway Crime Smarts; Hostage Situations; Jogging; Mugging; Murder; Push-Ins; Rape; Retail Store Crime; Self-Defense; Singles Bars.)

In 1958, I became one of the first two maximum security prisoners ever to escape from Folsom Prison. Folsom was built with Chinese labor in the 1800s, but despite its age it's considered one of the few escape-proof prisons in the world. From the outside, it looks like an impregnable fortress with walls 30 feet high and several feet thick.

Well, I can tell you that the prison's management didn't appreciate our record-breaking effort. My share of the reward was 4 years and 9 months in solitary confinement. When they recaptured us, I had someone watching me 24 hours a day. I was probably safer in the hole than in the mess hall. If you're unlucky enough to be having your lunch when a riot starts, something that happens a lot in prison, you can get shot just eating your soup.

Safe or not, most of us wouldn't want someone watching us 24 hours a day, even if the protection *was* free. If you're like me, you'd rather have some risk than face that much protection. How much protection you need depends on a number of factors: how you live, how much you're worth, your visibility, your job or profession, and so on.

HOW SAFE ARE YOU?

Here are a few test questions to ask yourself to figure out your degree of personal risk.

1. Do you live in a classy neighborhood and drive expensive cars?

2. Is your house or apartment in or near a city with a population over 100,000?

3. Are your parents wealthy?

4. Are you a key employee or an officer of a corporation with assets of more than $10 million?

5. Are you responsible for cash receipts over $5,000 on a regular basis (e.g., a cashier in a loan office, check-cashing service, or hotel)?

6. Do you make cash transactions in your work?

7. Do you live or work in a high-crime area?

8. Does your work involve making collections from customers or transporting valuable merchandise such as wholesale meats or liquor?

9. Do you drive a cab or bus in a large city?

10. Are you a sales executive who travels a lot, particularly to trouble spots like Central America or the Middle East?

11. Do you work late hours a lot or go out on the town at night?

12. Do you ride city buses and subways late at night?

13. Do you regularly travel on turnpikes, interstates, and big city freeways?

14. Are you a temperamental person who blows your top easily and gets into a lot of arguments and fights?

15. Do you believe crime reports are exaggerated and that most of us have little to fear from violent crime?

If you answered "no" to all of these questions, your risk is low. If you answered "yes" to one or two, your risk is average, but if you answered "yes" to more than two, your personal risk quotient is above average. And even if only one or two of them applies to you, you'd do well to take a look at ways to reduce these risks.

STRATEGIES FOR KEEPING SAFE

Uniform Crime Reports tells us a lot about crime in America. One thing they tell us plainly, in black and white, is that it's twice as dangerous to lead a normal life today as it was 10 years ago. Violent crime nearly doubled in the 1970s, and crime rates for robbery, assault, rape, and murder are near their all-time highs. Here are some things that can help you keep safe during a regular day.

Shopping

Shopping is not what it used to be. The risks are reduced because you can shop at many stores right from your own home or from where you work. Many stores let you shop by catalog or phone, and in some parts of the country you can shop for everything from food to clothes for yourself and the kids using cable TV shopping services. But for those of you who still go out to do your shopping, here are some tips:

1. Shop early and with someone. Plan shopping trips for early in the day. Coordinate your errands with those of a friend.

2. Be alert. Watch for deadbeats and loiterers in parking lots, parking garages, elevators, and walkways. If you see people hanging around, go another way. Report them to the store security as soon as you can.

3. Lock the car. When parking be sure to lock your car doors. It's a good idea to keep car doors locked at all times, regardless of whether you're in or out. When you return from shopping, check the interior of the car before getting behind the wheel, even if the car has been locked.

4. Avoid fights. Don't argue with other customers or store employees. If you have a problem, ask the clerk to call for the manager or a store security officer. If they can't handle the problem, call the police.

5. Carry little cash. Pay by charge card or check. Don't carry large amounts of cash if you can avoid it.

6. Have packages delivered. Whenever practical, have purchases delivered by the store or UPS. Don't load yourself down with packages that might interfere with your quick responses in an emergency.

At Work

1. Work in safe areas. Avoid jobs that take you into high-crime areas.

2. Travel with others. If your hours require late night travel, try to find another employee to travel with. Or arrange for family or friends to drop you off and pick you up.

3. Be alert. If you work in a large office building, be alert to suspicious-looking characters wandering the hallways, particularly after normal business hours. If you see messengers or other unauthorized people in areas where they shouldn't be, call building security immediately. Don't challenge them yourself.

4. Check out elevators. Don't get into elevators with people who look out of place or behave in a strange or threatening way. Report such people to building security or to the police.

5. Don't use the stairs alone. Stairwells can be traps as well as time-saving conveniences. Don't use the stairs unless accompanied by other employees. And whatever you do, don't enter stairwells to escape pursuers or potential attackers. Find an office with people in it and get help.

6. Lock up the car. Practice good security in employee parking areas. Always keep your car doors locked and windows rolled up. Don't even leave a little crack at the top for air. Even an amateur thief can easily open your car doors if you give him a break. If you notice anyone lurking near your car or in the parking area, report it to security and wait until they've checked him out before going to your car.

7. Be considerate and avoid confrontations. If disputes arise on the job or in business, refer them to a superior or third party for settlement. And remember that most disagreements can be avoided if you treat fellow employees and business associates with the same consideration, respect, and courtesy you expect from them.

8. Vary your routines. If you're a company officer or an executive with access to the firm's cash or vaults, you and your family are vulnerable to kidnapping attempts. Listen to one of the illustrations I give executives in my seminars on security.

If you come out of your house every morning at 8:30 and get into the same green Chevy, go two blocks and turn right, someday a kidnapper may be waiting there for you. You can throw kidnappers off the track by leaving for work and returning home at different times each day and by taking different routes.

Make sure everyone in your family understands the risks and practices good personal security. The danger of kidnapping is all the more reason to secure your home with good doors and locks and an alarm system. (See pages 16, 85, and 147).

At Play

1. Leave the sports talk at home. At the stadium or sports complex, don't discuss teams or athletes with other fans. You never know when a friendly discussion will hit some guy the wrong way. Don't take that chance. People can get hurt because of too much beer and too little smarts.

2. Be aware of your surroundings. Sports arenas, theaters, and nightclubs are often located in or attract predators from high-crime areas. So be on your guard when walking or driving in these areas. If no valet parking is available, look for supervised parking nearby. If you're forced to leave your car in an unsupervised parking lot or on the street, look for a brightly lit, conspicuous space, and make sure you lock up. Later, check your car carefully, inside and out, before getting back into it.

3. Take it easy on the sauce. You'll enjoy the evening just as much if you can remember it and can still manage to drag yourself out of bed in the morning.

4. Don't flash big bucks around. Whenever possible, use charge cards or traveler's checks and leave the cash in the bank.

5. Beware of bathrooms. You may think rest rooms are safe places, but when you drop your drawers don't drop your guard. If suspicious-looking types are hanging around, get out quickly. Try other facilities or wait till later. And do everyone else a favor and tell security or the police to check out the intruder.

6. Check out the taxi. Before getting into a cruising taxi cab make sure the cab company and driver are legitimate. Most cities require such cabs to be licensed, with the ID of the driver prominently displayed. Pass if you don't see both the name of the cab company and the driver's ID.

7. Let them have it. If you do get into a confrontation

with armed thieves, give them what they want without arguing. Then find a cop and let him go after them.

8. Avoid known trouble spots. This includes rock concerts. (See "Kids.")

At Home

1. Don't open the doors for strangers! This is the #1 prevention rule!

2. Keep your doors locked. Keep them locked even while you're upstairs, in the basement doing laundry, or washing dishes. Otherwise, you may have a surprise guest.

3. Secure your home with an alarm system. Buy a good perimeter alarm system and keep it activated when you're in the house as well as when you're out of it.

4. Install secure doors and good quality dead-bolt locks. Do this throughout your house or apartment.

5. Make your bedroom your "safe" room. Harden its doors and windows. If you have children, do the same for their rooms. Some people argue that the "safe" room concept is wrong because it puts you in a corner. I don't agree. If you wake up one night to find a 7-foot intruder standing over you, you won't either. You can make your "safe" room even safer by installing an extension phone and seeing to it that all other extensions are unplugged and safely locked away before you go to bed at night.

6. Communicate with neighbors. Set up a wireless intercom or some other signal system with your neighbor so that you can communicate in the event of an emergency.

7. Don't let them see you. If an intruder does get inside your house, avoid contact at all cost. If the phones are working, call the police; if they aren't, signal your neighbor and ask him or her to call the police. If you're out and arrive home to find someone has entered your house or apartment,

don't go in. Call the police from the nearest pay phone and
let them investigate.

On the Street

1. Face oncoming traffic. When walking city streets,
always use the sidewalk on the side of the street facing
oncoming traffic.

**2. Carry an electronic personal alarm in your purse
or pocket.** These are small battery-powered noise alarms that
emit very loud, high-pitched shrieks when activated. Just as
effective are the aerosol-powered pocket alarms that either
shriek or let out a bellow like a fire truck or fog horn. Both
can be bought at your local electronics outlet.

3. Keep your eyes and ears open. Look around you
as you walk. Know who's ahead of you as well as who's
behind you. If you don't like the looks of either, cross the
street. If they follow, start looking for help. Don't wait until
they're on top of you. And look out for suspicious types who
may be waiting in or behind parked cars.

4. Your best bet is to avoid high-crime areas. But if
you can't, be extra careful. Walk quickly. And stay away
from unpopulated areas and dark alleyways; they can be just
as dangerous as the high-crime areas if you happen to be
there at the wrong time.

Alertness and preventive action are your most effective
weapons against criminal assaults. You've got to recognize
the danger of living in today's society and then decide you're
going to deal with it. You can't expect the police to be solely
responsible for your protection. Each of us should do what
we can to reduce the chances of having crime happen to us.

Pickpockets
(See also Airport Crime; Credit Cards; Larceny/Theft; Personal Safety; Purse Snatchers.)

WHAT THEY ARE AND HOW THEY OPERATE

There are three kinds of pickpockets—dips, cutpurses, and cutpockets.

Dips

Dips are the ones you're most likely to run into. They operate on crowded city streets, on buses and subways, in theater lobbies, bars, and nightclubs. If you go to Atlantic City or Las Vegas, you may run into a dip in the form of a *chip snatcher*. This specialist lifts valuable casino chips from the rail of the table in front of you—or from your coat pocket or purse.

Dips come in a wide range of age groups and skills, from clumsy kids and addicts to top pros who can lift your wallet, take your money out of it, and put it back in your breast pocket without your feeling a tickle. But there aren't many of these pros left. Most of today's dips just take your wallet, charge cards and all, and drop back into the crowd.

Cutpurses

Cutpurses earned their place in history in post-Elizabethan London where the local dandies carried their guineas in charming little purses pinned to their doublets. The cutpurses would simply slit the bottom of the purse with a sharp dagger and let the coins fall into his hands. His modern counterpart cuts the strap of your purse and takes off with the whole thing. (See also "Purse Snatchers.")

Cutpockets

These soft-touch artists are probably the rarest type of pickpockets. For obvious reasons, his victims tend to be prosperous-looking gentlemen he meets on the street, in a crowd, or wherever.

A friend of mine who was in Washington recently told me about a funny spectacle he observed there. One of our populist lawmakers, who takes special pride in his daily contacts with the masses, came out of the subway with his morning paper firmly underarm. As the Congressman strode purposefully off, my friend and the rest of the crowd couldn't help noticing that a patch of custom suiting material a little larger than a wallet had been judiciously removed, as they say, from the worthy pol's pants, revealing shocking pink undershorts.

A good cutpocket can take a pocket out of your coat or trousers in the time it takes you to turn the page of your morning newspaper on a crowded, rush-hour train or bus—and probably with less trouble. The hell of it is that besides losing your cash and charge cards, your favorite suit gets ruined.

WHAT YOU CAN DO TO AVOID PICKPOCKETS

Your chances of having your pockets picked are actually quite low. Pocket picking accounted for about 1% of larceny in 1982. If you take ordinary precautions, you can easily protect yourself from pickpockets. Today's pickpockets just can't find time to practice, and practice is what it takes to be a good dip.

Here are a few tips to follow to minimize the risk even more:

1. Keep it closed. Don't carry your wallet in a purse without a clasp or zipper closure. And make sure you keep your purse closed at all times. Open purses are a cinch, even for the clumsiest dip.

2. Keep it out of sight. If you use a card case or wallet, keep it securely tucked away in your purse or the breast pocket of your suit or jacket. Don't carry a wallet in a hip pocket unless you have no other choice; then be sure to keep the pocket flap or button on the pocket securely fastened.

3. Keep it small. Don't carry a long, full-sized billfold. Look for a small folding one that will disappear in your pocket or purse. Long billfolds are great for impressing headwaiters, and dips love them too.

4. Keep it hidden. In crowds, try not to let people know where you carry your wallet. The first thing a dip does when he spots a likely candidate is try to find out where he carries his stash. Don't give him any clues by handling or displaying your wallet unnecessarily.

5. Keep them separate. Carry your money and charge cards in separate places.

6. Use a money clip. A money clip is a good, safe way to carry paper money. Keep it in a pocket or someplace other than in your purse or wallet.

7. Use a money belt. If your work requires you to carry large amounts of currency in crowded, high-crime areas, use a money belt.

8. Be alert. Again, as I remind you throughout this book, when it comes to crime prevention, there's no substitute for good, old-fashioned alertness. That's especially important in crowds. When you start feeling bodies around you, prepare yourself for attempts by dips. Don't let people jostle or bump you without a defensive response. Grab your

176

watch and your wallet, and make sure they stay where they belong.

9. Carry an alarm. If you live or work where there's an epidemic of pickpockets, there is one bit of technology especially designed for you. It's a pocket-size, battery-powered, electronic personal alarm with a short braided-metal chain securely attached to a plastic card that you insert in your wallet. If anyone tries to meddle with it or to remove the wallet from your pocket or purse, it sets off the alarm with a shriek you can hear for half a mile. These alarms are available at your local discount radio/TV outlets and where electronics equipment is sold.

WHAT TO DO IF YOUR POCKETS ARE PICKED

1. Report it to the cops. Don't bother to question the people around you or to try to flush out the pickpocket. Most likely you didn't even feel it when your wallet went, which makes it difficult to pinpoint your location when it was stolen. Go immediately to the nearest police station and file a stolen goods report, describing your wallet, its contents, and where and when you think it may have been stolen.

2. Report it to your insurance company. If your insurance covers out-of-home losses, report the theft in writing to the insurance carrier. Include a copy of the police report.

3. Report your loss. If you were carrying charge cards in your wallet, inform the issuing companies both by telephone and in writing, immediately upon discovering the loss. (See also "Credit Cards.) Be sure to report if any other items such as a passport, Social Security card, or birth certificate were stolen, so that the police can inform immigration officials and other federal agencies.

4. Keep an eye out. Watch for suspicious strangers hanging around your home or apartment. If the pickpocket likes what he finds in your wallet, he may take a notion to come around for the rest. (See also "Burglary," "Armed Robbery," and "Personal Safety.")

Purse Snatchers
(See also Mugging; Pickpockets.)

The law says that a guy who grabs your purse is committing a nonviolent property crime. I say look out for that turkey!

In uniform crime reports, purse snatching is lumped together with picking pockets and bicycle theft. But the truth is, purse snatchers often use physical force to separate you from your purse, which makes it *assault* and *robbery*. In what appears to be a trend toward unnecessary violence, more and more women are being hurt in these encounters.

Don't try to fight violent crime with violence. So forget about taking those kung fu lessons from the local hunk. Your efforts to prevent your own purse from being snatched should be based on good old common sense and should square with your other precautions against crime. Stay alert; harden the target; and don't take risks—particularly against unknown odds.

Yes, purse snatchers usually *are* punks and small-time crooks. It's a common crime among teens on drugs looking for money for a fix, street people who'd rather steal than work, excons who can't get a job, and problem kids or adult low-life types. But don't get the idea that purse snatching is a harmless, infrequent occurrence. Crime records show that purse snatching was up 10% between 1978 and 1982, which indicates that it is a definite problem you have to watch out for.

HOW TO DISCOURAGE PURSE SNATCHERS

1. **Remove the opportunity.** My guess is that a lot of women carry a purse out of habit rather than because they really need it. You carry charge cards instead of a checkbook and cash. You often wear pantsuits, jeans, and jackets instead of skirts and dresses, which means a lot more pockets to carry things in. So why carry a purse if it's going to increase your risk from predators—and believe me, it is. This is particularly true in high-crime areas. If you must go into them, use your pockets and leave your purse at home.

2. **Carry small.** If you must carry a purse, carry the smallest one possible. Try to find a small, slippery one (the more it resembles a N.Y. Jets' football, the better). Forget the shoulder strap. The purse-snatchers favorite target is the bag with the long, generous shoulder strap—one that he can get a good grip on before yanking you off your feet and disappearing into the crowd with your worldly goods. He also likes these bags because they're the ones he finds hanging on the backs of chairs in restaurants and other public places. To really throw a purse snatcher off the track, carry your purse in a brown paper bag.

3. **Keep it close.** Carry your purse under your arm or in your hand against your body. Keep the closure next to your body. Don't drape it over your shoulder and/or behind you. Some guy can grab it and pull you over backward before you know what's happening. Or, if he's a hotshot with a straight razor, he may just slit it open and help himself while you strut your stuff.

4. **Hold on to it.** Never leave your purse on the floor at your feet or hang it from coat hooks, on the backs of chairs, or on barstools. Keep it in full view, either on top of the table or in your lap.

5. **Carry an alarm.** If you must carry a large bag,

there is a special battery-powered electronic alarm designed to use with a purse or wallet. This little gadget won't stop the guy from grabbing your purse, but it will make him so miserable when he does that he's likely to let go in a hurry. Like the other pocket alarms I recommend, this one lets out a shriek you can hear for half a mile. Since its armed by punching in a digital code, only you can turn it off. It's made of unbreakable plastic and is literally indestructible. A purse snatcher won't want to hang on to anything making that much noise for very long. You can buy these little alarms from your local electronics outlet for as little as $12 (including tax). Just ask for a wallet/purse personal alarm.

6. **Stay alert.** This is particularly true when you're in neighborhoods known for frequent street crime. Purse snatchers like to catch you off guard or preoccupied. If they see that you're alert and aware of them, they'll look for someone they *can* surprise.

7. **Don't overload yourself with parcels or briefcases.** If you're shopping, have things delivered. This is particularly important during the Christmas season, because purse snatchers (and criminals in general) get the Christmas bug, too. They're on the lookout for ladies juggling armfuls of parcels as well as their purses. The same goes for men who are so loaded down they can't move fast.

WHAT TO DO IF A PURSE SNATCHER STRIKES

1. **Don't lose your head.** If your purse gets snatched, ask yourself a few questions before you make your next move:

- Did anyone witness the crime and move to help?

- Is the purse snatcher sober and rational?

- Did he threaten you with violence or show a weapon?

- How important or valuable are the contents of your purse?

Your purse snatcher *may* be a sneak, a coward, and an emotional basket case who will dissolve into tears at your first scream. But if he's not, and there's always a good chance that he isn't, you *could* be in more trouble than what's inside your purse is worth. So, don't take off after him.

2. Let it go. If your purse is a shoulder bag and someone grabs it, let it go. Don't get tangled up in it or let the snatcher drag you along with it. Many of the injuries reported happen when 200-pound purse snatchers get tired of struggling with 100-pound victims.

3. Scream. Start screaming at the top of your voice as soon as you're out of reach of the snatcher. Make as much of a scene as possible, but don't run after the purse snatcher. That's what you pay the police to do.

4. Call the cops. Go immediately to the nearest phone and call the police or the designated emergency number in your area. Give them a detailed physical description, including age, sex, clothing, height, weight, coloring, hair style, visible marks, and scars, as well as the direction of escape and anything else that might lead to the arrest of the purse snatcher.

5. Visit the precinct. Follow up your call with a personal visit to the nearest precinct house or police station. File a formal complaint with full description of your purse and contents, and get a copy of the complaint for your insurance company.

6. Report missing keys. Be sure to let the police know if your purse contained keys and ID. They may want to add patrols in your neighborhood, because criminals have been known to use keys and ID from purse snatchings to commit further offenses such as burglary and robbery.

7. Change your locks. As soon as possible, have the cylinders changed in any locks whose keys were taken with your purse, including the locks on your car.

8. Call the card companies. Notify the issuing companies for the charge cards stolen with your purse. (See also "Credit Cards.") Also, advise your bank if your checkbook or passbook was stolen.

9. Alert your friends. This may sound extreme, but you should make sure anyone listed in a stolen address book is alerted to the theft. Crooks have been known to use these names to run scams, setting people up for swindles, burglaries, or robberies. Urge your friends to be suspicious of all strange callers.

10. Tighten your security. If you've been putting it off, now's the time to increase the security of your home or apartment. Upgrade locks (see "Locks") and install a perimeter alarm and outdoor lighting (see "Alarms, Home and Business" and "Burglary") to protect against any further theft by the purse snatcher.

11. Don't play the heroine. If you should meet your purse snatcher on the street, don't take revenge into your own hands. Call the police. (See also "Robbery" and "Personal Safety.")

Push-Ins
(See also Armed Robbery; Elderly, Common Crimes Against the; Mugging; Personal Safety; Rape.)

Push-ins are usually one step in the enactment of other violent crime such as armed robbery, assault, and rape. But because of the increasing frequency of push-ins, it deserves special attention.

Push-ins are big city crimes, and older citizens and lone women are the primary victims. They usually occur in one of two ways. In the first, you are surprised from behind as you enter your house or apartment, shoved through the door by your assailant, and robbed in the privacy of your own home. In the second, you are already in the safety of your home, respond to the doorbell, and are confronted by an armed intruder who pushes his way in.

COMMONSENSE WAYS TO AVOID PUSH-INS

Push-ins are a violent crime. They're dangerous from the moment they start. Criminals who practice this crime are responsible for a disproportionate share of violence. For that reason, it's important that you learn how to prevent them. They're a lot less traumatic to prevent than to deal with when they're happening.

If you are a single woman, a woman who's often alone in the city, or an older person living in the city, you know push-ins are a very real possibility. Keep that thought with you wherever you go—and learn to take the following precautions:

1. **Make sure the lights work.** Many push-ins happen after dark when the victims are returning home. *If you live in an apartment,* make sure the building management keeps entries and hallways brightly lit. Don't go in if the lights are out. Sure, the bulb may have burned itself out, but removing bulbs is a favorite trick of robbers. Find a pay phone and call the manager or super, or get the cop on the beat or a friend to escort you to your door.

If you live in a house, keep the shrubbery around the entrance trimmed back and the driveway, walks, and entrances well lighted. Put your outdoor lights and some indoor lights on automatic timers set to turn on before dark.

2. Keep a sharp eye out for anyone or anything that looks fishy. Your alertness and ability to spot trouble before it happens is your best defense against push-ins. Don't lull yourself into a false sense of security by telling yourself it won't happen to you. Raise your level of paranoia and be suspicious of strangers. If a stranger follows you into your apartment building lobby, elevator, or stairway, let him go ahead. Don't go in if you see anything that might represent a threat. Stroll around the block, visit a neighbor, or chat with your favorite waitress or bartender. Later, if things still don't look right, call the police. Have them check things out before you put the key in the door of your apartment.

3. Keep an eye on parked cars. If your entrance is close to the street, watch for people sitting in parked cars. This is a common ploy of characters who don't want to be seen loitering in the street. If someone makes a move to get out and follow you, don't go in. Go to the nearest phone and call the police. Have them check out your mystery visitors.

4. Be suspicious of any unfamiliar face in your building. If strangers board the elevator with you, get off at the next floor and take the next available elevator car to the lobby. *Don't lead possible intruders to your apartment door.* Get the manager, or the super, or someone from building security to ride back up with you. Don't take chances with strangers wandering around the hallways.

5. Keep that door closed. I say it again and again throughout this book, because it is one of the most important requirements in home security. Don't open your door to anyone you don't know, aren't expecting, or can't identify. Make sure your door is equipped with a wide-angle optical viewer (peepholes). Don't unlock your door unless you can identify your caller.

IF YOU ARE PUSHED-IN

1. Keep your head. Don't waste valuable time struggling with someone already in your door. If you're not being threatened with deadly force and you have a burglar alarm, set it off if you can. If your apartment window faces the street, pitch a heavy book or some other large item through it to attract attention. Or light a matchbook and drop it in your wastebasket to get a fire going. If you've followed my advice and equipped yourself with a wireless intercom, give your neighbor a holler. Keep things moving for the intruder, and do your best to attract attention before the crook has a chance to establish his position.

2. Remain calm. If you *are* being threatened with deadly force, particularly a knife or gun, try to stay calm. Do what he tells you to. Find out what he's after and tell him you'll give it to him, if you can. Don't waste time and endanger your safety by trying to bargain or reason with your intruder. Remember, once he's in, your task is to get him out as quickly as possible.

3. Call the cops. When your push-in has taken what he came for and left, call the police. Don't be intimidated into remaining silent. Give the police a full description. As crooks often work particular neighborhoods (for them, it's the next best thing to living there) there's a good chance you may catch sight of him again. If you do, call the police so that they can pick him up, and get him off the street.

R

Rape
(See also Computer Dating Services; Hostage Situations; Mugging; Personal Safety; Push-Ins; Real Estate Risks; Singles Bars.)

Forcible rape and murder are the two offenses considered to be capital crimes by every society in history. In some non-Western cultures, rape is punishable by death, and its victims are ostracized. In the U.S., forcible rape is a crime of violence against the state as well as against the individual.

YOUR CHANCES OF BEING RAPED

Over 77,000 women were raped in the U.S. in 1982. Considering that not all forcible rapes are reported, this suggests that if you're female and 15 years or older, your chances of being raped this year are about 1 in 1,000. The likelihood of your becoming a victim of rape depends on things like where you live, what you do for a living, and how you spend your free time. For example,

• You're over twice as likely to be raped if you live in the Sunbelt than if you live in the North or East.

• Your chances of being raped are much greater if you live in or near a city with 100,000 people or more.

• You're almost twice as likely to be raped if you're black.

• Your risk of rape is greatest if you're between 15 and 35.

• Even the seasons affect the chances of being raped: The risk is greater in the summer than in the winter.

Between 1973 and 1982, the incidence of forcible rapes increased by 51%. Rape has become truly epidemic, largely, I believe, as a result of the inability of the police and the courts to prosecute and sentence rapists and sexual offenders effectively.

6 IMPORTANT STRATEGIES FOR AVOIDING RAPE

By far the largest number of forcible rapes are premeditated, in the sense that the rapist selects his victim ahead of time. Although spontaneous acts of rape do occur, they're rare. The usual scenario goes something like this: The rapist spots his victim, follows her, waits for the right opportunity, then strikes.

The situation sounds grim, but there's good news, too: rapes *can* be avoided. Here are a few strategies for protecting yourself against rape:

1. Face the facts. First and most importantly, convince yourself of the fact that *all* women are potential rape victims. Facing up to your vulnerability is the most important factor in rape prevention.

2. Keep your guard up. You may have to tighten up a little in the way you do things. Keep home and car doors locked. Always check the back seat of your car, even if it's been locked, before getting into it. Keep your eyes open while shopping or walking in public places, parking lots, etc. Take a friend along, if possible. Don't overload yourself with packages that block your view and make it difficult for you to move in case of an emergency.

3. Keep your guard up. Don't be lulled by ordinary situations. Whether you're out shopping, on your way home from work, or just walking the dog, keep your paranoia up. Don't trust anyone: The kid running the lawnmower or the guy driving the pickup could be a threat. The most ordinary situation can be the most dangerous, simply because you don't expect it.

4. Use your common sense. Don't adopt the "free spirit" attitude that will lead you to do impulsive or dumb things that your common sense tells you are wrong. If the urge hits you to jog alone or to take walks in isolated or potentially dangerous areas, stifle it. And the same goes for using basement or public laundromats late at night, visiting the roof of your apartment house, and using rear stairwells by yourself. Anyone who's stalking you will welcome these invitations.

5. Be cautious. Make it a point to be reserved and cautious with men you don't know well, in professional as well as private situations. Don't flirt with or tease strangers, unless you mean business and are ready to accept the consequences. Body language may sound chic and trendy coming from some pop psychologists, but you should be aware of the possible misunderstandings that can result from certain kinds of signals. And whatever you do, don't hitchhike or accept rides from strange men.

6. Develop good personal safety habits. (See "Personal Safety.") If you think you're being followed, don't panic. Change sides of the street and stop in one or two open shops or bars. Once you're sure you're being followed, try to get as much of a description as possible without his catching on.

189

Then get to a pay phone and call the police. When they arrive, give them the guy's description and tell them you want to file a John Doe complaint. Make sure they get it on the radio and are actively looking for the guy before you leave the station. If you see the guy's face again, get on the horn. Don't let him follow you. Under no circumstances should you run into empty offices, industrial buildings, stairwells, or elevators; or try shortcuts through parking lots; or hide in alleys or behind parked cars in an effort to get rid of the pursuer. Find a crowded place where you can use the phone, or stay in the open street where you have as many options as possible. In an extreme emergency, where time is becoming a critical factor, try to find an unlocked car, get in, and lock all the doors and don't open them for anyone but the police.

WHAT TO DO IF YOU ARE RAPED

Rape is an outrageous violation that can be both psychically and physically damaging. But the damage may be healed with time, unlike murder. So personally, I do not recommend that women carry guns or learn kung fu for use against assailants. I've met some of those turkeys who've raped women, and I can vouch for the fact that too many of them are better at that stuff than you could ever hope to be. They've been shot and stabbed more times than you've had birthdays, and they just keep coming back for more. That can be real discouraging when you've run out of bullets for your gun.

Here are some things to do if, however careful you are, you are raped:

1. Get away from your attacker as quickly and safely as you can. Find a public place where other people, preferably women, are available to help you. If you're out in the country, look for a hospitable-looking house or an open service station.

2. Call the police; give them details of the attack and a description of your attacker.

3. Call a friend or family member who can come and be with you.

4. Preserve all physical evidence. Don't shower or bathe or otherwise clean up before your examination. Save the clothing you were wearing. Show police the site where the assault occurred.

5. Go to the nearest emergency hospital for medical treatment and examination. Make sure they understand you're a rape victim and that they need to preserve the evidence.

6. Call your local community service information office and ask for the telephone number of a counselor at the Rape Crisis Center or a similar organization. If you're a rape victim, don't let the experience isolate you. You're not alone. Women who have suffered the same kind of violation are usually available to offer support.

Real Estate Risks
(See also Armed Robbery; Larceny/ Theft; Rape; Personal Safety.)

When we put our homes, condos or co-ops up for sale—on the market, as they say—our minds are likely to be on the details of swinging a deal and what we're going to buy next. We aren't thinking about the risks. But they are there regardless. And you need to be aware of them so that you can guard against them if you are selling your home.

WHAT ARE THE RISKS?

Theft

Not long ago, some friends of mine listed their home with a prestigious realtor in an affluent suburb of a large city. Shortly thereafter, one of the realtors' salespersons showed up with a client, a woman she described as a well-to-do psychiatrist who lived in the community but said she was looking for a bigger house.

After the psychiatrist and the realtor left, my friends noticed that several small items were missing, including an heirloom, sterling-silver bar measure. Later, my friends took their house off the market. They still live there, minus their silver bar measure.

The moral of the story is if you or your agent plan to show your furnished home, be sure you secure your valuables. You never know when someone with sticky fingers is going to walk in the door.

Violent Crime and Burglary

One of the biggest reasons for paying a real estate broker a commission is his or her ability to check out prospective buyers in advance. The theory is if the agent doesn't like the customer, he or she can send the person down the street.

Not anymore. The legislators have legislated our ability to protect ourselves out of existence. Today, the agent is legally bound to show a property to anyone who asks to see it, though he might be a rapist, murderer, or Dr. Frankenstein's monster. Here's a look at some of the consequences of the new legislation:

1. People are being robbed by crooks posing as house hunters. These phony clients appear with the agent, pull guns, tie everyone up, take wallets, watches, and jewelry,

toss the house, then back up the truck and move your stuff out. By the time one of you gets loose, they're counting their profits.

2. Women real estate agents and in some cases wives and mothers showing homes or apartments during the day are taken into bedrooms and raped.

3. Professional burglars, posing as very fussy prospective buyers, poke into every corner, checking alarm systems and other forms of security, and noting the types of locks, the layout of rooms, the location of a safe, etc. Later, when you've gone shopping or to work, they clean the place out in record time.

HOW TO MINIMIZE REAL ESTATE RISKS

You'll want to protect yourself as well as your broker's salespeople from robbery, rape and burglary. Here's how:

If you've decided to sell your house or apartment, make sure each prospect has been carefully checked out before letting him in the door. Don't worry about insulting a prospect. People serious about buying will understand that they will have to give this kind of information to mortgagees or other lenders sooner or later and will have nothing to hide. Anyone who gets all bent out of shape is probably a phony who's just wasting your time—and may even be one of the crooks I'm talking about. Here's a list of questions you or your broker should ask potential buyers:

- Full name

- Current address and phone number

- Present living situation—own or rent?

- Name and address of mortgagee or landlord

- Name and phone number of business or employer

- Name of bank

- Three business references

- Three personal references

Only after you've checked out all references should you or your broker make an appointment for a showing.

Retail Store Crime
(See also Alarms, Home & Business; Armed Robbery; Burglary; Safes; Self-Defense.)

Unlike other crimes that affect specific groups or individuals, retail store crime hurts us all. We all have to buy the food and clothing sold in retail stores, whose prices are forced up by losses.

Retail associations estimate 1983 losses from crime will be close to $7 billion. FBI reports show that larceny/theft offenses are up 21% since 1977. Combined, these figures indicate that retail store crime has increased by over $1 billion a year. Can you and your children really afford to live with these crime rates for the rest of your lives? I don't think so. I don't think any of us can.

Okay, but what can you, as a customer, do about it? The fact is, you can do plenty!

1. **Fight the urge to get even.** You can fight down the urge to get even for the higher prices by slipping that extra package of hamburger under your copy of *Family Circle*.

2. **You can keep your eyes open.** If you see anybody else trying to get away without paying for something, you can turn that bimbo in. Why should you pay more so he or she can have a free ride?

3. You can do a little pioneering with your kids. Most likely your kids don't shoplift, but an awful lot of school-age kids do. In fact, in some crowds, they even make a game out of it and keep score to see who can steal the most. Many are middle-class and upper-class kids who don't need the stuff; they do it for kicks and because they know they won't be prosecuted even if they get caught. Kids actually account for a large number of shoplifters. They probably don't make the connection between what they take and why its price keeps going up. Tell them about it, and get them to talk to their friends.

CHECKLIST FOR STORE OWNERS

If you're a store owner or manager, you should have three crime prevention priorities: robbery, because it's dangerous as well as costly; and burglary and shoplifting, because they've become major factors in the price increases that hurt sales.

Those are the priorities we work with at 7-Eleven, the largest convenience store chain in the world. As you read in the forward to this book, since I began working with the 7-Eleven stores, we've made good progress in reducing these crimes through prevention. You can do it too. Here's how:

1. Reduce your robbery appeal. Deep-six all the posters, broadsides, and banners in your windows and wash the windows so everybody in the street can see inside the store. Then upgrade your interior lighting so every corner of the store can be seen from outside. Put your safe right in front of the window. Robbers and safecrackers hate audiences. They'll look for someplace more private.

2. Use cash controllers. If you're open late or during odd hours, install a cash controller like the ones we use at 7-Eleven. Controllers give us a safe place to get rid of big bills and provide controlled amounts of cash during the shift for making change. The people who install the controller will put up a sign for you, telling the crooks you have a new gimmick for them. The experienced robbers will pass; the idiots could wind up doing one to ten for $35.

3.   What about store surveillance? Are you using those big, round mirrors to watch your aisles? If you are, you probably don't see much. Besides, the guy on the other end can see you just as well. He simply waits until you get busy, then makes a move.

4.   Start cutting your losses, so you can stop raising prices. Take a hard look at the way you put out stock. Can you rearrange things so that the most often stolen stuff is less accessible to shoplifters?

5.   Install an alarm. Get a good quality perimeter alarm system, and make sure your doors and windows are force resistant and have good locks.

6.   Get a dog. Leave your dog in the store at night.

7.   Prosecute. If you do catch a teenager shoplifting or a burglar, prosecute to the fullest. Put up signs around your store warning shoplifters and others that you will prosecute. Don't back off if one of them turns out to be the mayor's nephew.

8.   Notify your customers. In your advertising and promotion, tell your customers what you're doing to fight crime. Explain to them the annual cost of crime to people in your business, and offer to lower prices when there's a substantial reduction in the crime. Get them on your side and working with you. Try to make crime prevention a community crusade.

S

Safes
(See also Alarms, Home and Business; Burglary; Apartment Security; Guns.)

Safecracking was never my strong suit. But I have friends who knew that game well. One of them, a man I've been proud to know for many years, is a former safecracker. He doesn't like the term because, as he says, he doesn't *crack* safes. He just opens them, using superior knowledge and skills. As a young guy, Tom was one of the best safe men in the country. He got started wrong but turned his life around and became very successful in the straight world. Today, he runs a well-established, highly respected security consulting business headquartered in Dallas, Texas, that specializes in the installation and servicing of safes and vaults. Major companies in the Southwest have security systems designed and installed by Tom's company. He's one of the best in the business.

This is why I've asked Tom to give us some answers to common questions about safes:

PROS AND CONS ON SAFES

RAY: Tom, what are the pros and cons of having a safe installed in your house or apartment?

TOM: I can't think of any cons, Ray. Everyone ought to have some sort of safe. Your readers ought to be cautioned about getting too much or too little safe. They should keep in mind what they want to store in it. In some cases, a good, heavy steel box bolted to a large chest or workbench is as good a safe as you need. Safes and vaults can be expensive and there's no use spending a small fortune on something you're going to be keeping your extra set of dentures in.

RAY: That reminds me of the time my buddy and I burglarized a business in San Diego. The two of us wrestled a 500-pound safe into a Ford two-door sedan and hauled it out into the boonies so we could work on it in private. No matter how I turned the safe, it wouldn't come out of the back seat it was about two inches too big. I worked on the safe for about 12 hours—sweating and swearing, and straining. Finally, I cracked it. When I looked inside, I couldn't believe my eyes. The sole contents of that mother was 60¢ in postage stamps and half a cheese sandwich!

TOM: That's right. And that safe sounds like it probably cost somebody about three grand. It was a lot more protection than that sandwich deserved.

RAY: Okay, then, what are some of the pros?

TOM: Safes are useful to almost everyone. They don't have to be expensive or elaborate. If you're not into collecting gold coins and diamonds, you don't need to spend a lot. You can pick one up from your local department store for under $500. Properly installed, it will protect you from all but top-notch pros. And unless you're keeping a lot of interesting stuff in it, you'll probably never attract a profes-

sional safecracker. These lower-priced safes provide a secure place for credit cards, securities, spare cash, family
documents, as well as things you don't want the kids to get
at like registered handguns, prescription drugs, and liquor.
Sometimes a simple metal cabinet with a good lock on it is
all you need. But everyone should have something the
neighborhood kids can't break into.

RAY: Tom, what specific tips can you give my readers about
the best safe to buy, how to protect it, and so on?

TIPS FROM A SOMETIME SAFECRACKER

Here's Tom's advice about safes:

1. Determine how much safe you need. For the average
homeowner or apartment dweller, a fire-retardant metal cabinet will do. A lockable lateral or vertical filing cabinet can
fill the bill nicely. The lateral steel files with two drawers can
also be used to hold rifles and shotguns if you break them
down, and they provide plenty of room for liquor and other
large items. You can buy these files new from $50 to $100.
Fully loaded, they're clumsy and difficult to move; they have
the added security advantage of being easy to bolt to wall
joists or floors.

2. Know your options—and use them. If you live in a
rental apartment, your options are limited. Experienced burglars know most of the places you can hide a portable safe
such as inside tables or other furniture or installed in walls
behind bookcases, paintings, etc. Unless you're able to set
them in concrete, any of these small safes can be pried loose
and carried off to be opened later. They may offer enough
protection against kids and amateurs, but not with pros. If
you live in an apartment house that attracts a lot of burglars,
your best bet is to use a safe deposit box at the bank.

3. Floor safes are good. If you live in a single family

residence, co-op, or condominium with a basement or slab floor, you're in luck. Floor safes that can be set in concrete are available. To hide yours from prying eyes, keep something like the TV set over it. A good floor safe, properly set in concrete, is virtually indestructible. The only way in is through the combination.

4. Keep it visible in public places. If you're installing a safe in a business, a store, or a service station, install your safe at the front of the store. The point is that no crook is going to stand around waving a gun or pounding on a safe with a sledge hammer in front of customers, passing motorists, and patrol cars.

5. Make sure the safe's rated. If you decide you need something more than just your everyday small residential safe and believe your needs justify the additional cost, make sure the safe you select is rated. Ratings by Underwriters Labs, Inc., are coded and stamped on the manufacturer's name plate or somewhere else on the safe. Codes indicate the rated strength of the safe. They usually begin with the letters "TR" (torch resistant) or "TL" (tool resistant); better safes will have both. They are followed by numerals—10, 20, 30, etc.—representing the number of minutes it would take an expert, using the latest available technology in torches or tools, to open the safe. An additional code—the letters "A," "B," "C," etc.—represents the fire-resistant capabilities of the safe—"A" representing 5 minutes, "B" 10 minutes, "C" 30 minutes, "D" 60 minutes, and so on—to 1700 degrees of flame.

6. Install small safes in concrete. Small safes that weigh less than several hundred pounds should be installed in poured concrete forms, which add a minimum of 200 pounds to their weight, and the entire form and safe should be firmly anchored to a reinforced wall or floor. If you're installing it in cured concrete, such as in a slab or basement floor of older construction, take out enough old concrete to make a hole large enough to allow for at least 6 inches of new concrete on all sides of the safe. Make sure the old concrete is thoroughly wet before pouring the new.

7. Change the combination. Once your safe has been installed, have the combination changed by a safe expert you can trust. Otherwise, you'll wind up with the same combination as everyone else who takes delivery in your area. Each manufacturer of safes has what is called an international shipping code, or combination. Safes destined for the same market all have the same code. This is done so that the people who sell the safes won't receive safes that they can't open. You are sold the safe with the code still on the dial. Any locksmith will have a list of these codes going back for years.

8. Avoid anything electronic. We're on a *Star Wars* kick right now, with our cars, our tools, and everything else going digital. In a few years, we're going to regret it, when we have to pay to fix all these things. Most of the stuff doesn't work a damn bit better than the simpler things they replace. Electronic safes give you all kinds of problems you don't need, like power failures, backup systems malfunction, diode failure, and component breakdowns of other kinds, to name only a few of the most common ones. There may come a time when all these little resistors and capacitors will last forever. For now, you're better off with an old-fashioned dial combination.

9. Hire someone trustworthy. Be careful who you hire to change the combination on your safe, both at the time you have it installed and later. If you live in a small community where everybody knows everybody else, there's no problem. In a larger community, don't just pull a name out of the phone book under Safes & Vaults. Check with the Better Business Bureau, the guy who handles your account at your bank, or someone else who can recommend someone who's been around awhile. There's no use spending a bunch of money on a safe if the information on how to open it isn't going to be secure.

10. Keep the combination secret. Once you have the new combination, don't write it down or give it to others, especially youngsters or others who might inadvertently reveal it. Don't use your birth date, anniversary, or license number as the basis for your new combination. Make sure the number is one that no one, not even the people who know you well or have made it their business to study you, will be able to hit upon. Again, a good safe is a big investment—and the security of the safe boils down to the security of the combination.

11. How much should you spend on a safe? That's easy. How much would you pay to get your stuff back if it were stolen? In my business, I sometimes get calls from people who want to buy a safe for their home or business. When I tell them the cost, they decide against it. Later, they're burglarized. Then I get a call from them to order a safe, and this time cost is no object; they need the protection. It's human. But they would have saved themselves a lot of pain and expense if they'd made the decision earlier. If you think you need a safe, you probably do. The question you have to answer is one only *you* can answer. How much safe do you need?

12. Call a professional. If your safe has been tampered with, call a professional before messing with it—the door could fall off and break your foot—or some nut who watched too much TV might have tried to open the safe with an explosive which could still be active.

Self-Defense
(See also Guns; Personal Safety; Rape.)

This whole book is about self-defense. It's about *prevention* as self-defense, and it's about *avoidance* as self-defense. I don't talk much about *reacting* in self-defense, because I've got this theory that if you have to react to something, it's already too late.

To a lot of people I know, self-defense is fighting to keep some cat off of you by getting at him. It's violent confrontation. And in the world we live in, that's very dangerous.

Perhaps, if I hadn't lived for so many years in prison, I'd look at it differently. In the joint, I learned the difference between the way people who *care* about life and living are violent, and the way people who *don't care* are violent. Too many of the people I knew in prison were in the latter category—they didn't care, and they had nothing to lose. When you're like that, you're violent in a whole different way from when you're thinking about life, your job, your wife, the kids, and so on.

My point is very simple. When it comes to violence, people who care are no match for people who don't. That often translates into this: When it comes to self-defense, honest people are no match for criminals.

That's not to say you shouldn't defend yourself if you're good at it. For all I know, you may be another James Bond. But this book is written for *everybody*, and not everybody is up to taking on reckless criminals. No one I've ever known has successfully smashed a tire iron with a karate chop.

My advice, which I repeat again and again, is to *avoid* violent confrontation. But I've included this section in self-defense because there are times when you can't. Here are a few tips that may help keep you healthy and your friends and family safe.

GUNS, KNIVES, CLUBS, AND OTHER WEAPONS

Guns

As I pointed out in the section, ''Guns,'' the records clearly show that guns are far more dangerous to their owners and the people around them than they are to criminals. There are a lot of situations in life where we can afford to be less than perfect. Playing with guns isn't one of them. Guns are very unforgiving. You only need to make one mistake.

If you feel you need a gun for self-defense, you owe it to yourself and others to learn as much as possible about the safe use and security of guns. If you aren't already an expert, make every effort to become one. Before you arm yourself, write or call your local chapter of the National Rifle Association, or check with your local sporting goods store or gun shop for the name and address of a reputable gun club. Become a member and follow their instructions.

Knives

For my money, knives are not a defensive weapon at all. People who use knives in the street aren't likely to give you a warning. The only thing a switchblade or spring-blade knife can do for you is get you into more trouble than you're already in.

Clubs and Other Weapons

Clubs and blunt instruments—including ax handles, baseball bats, taped pipe, shot bags and saps, billy clubs, blackjacks, and chains—all share the same drawback: None are bullet proof. If you try to use one on an armed criminal, all it can do is get you shot. If you use one on somebody who isn't armed, you'll need a good reason. Nailing somebody because he runs into you on the freeway probably won't be a good enough reason for the judge, and you could wind up in trouble yourself.

COMMONSENSE SELF-DEFENSE

1. Run like hell.
2. Carry a personal alarm or aerosol power horn to frighten off would-be attackers. (See "Personal Safety.")
3. If you don't have an alarm, scream a lot.
4. Do not, repeat, do not attend judo or karate classes with the idea that, boy, the next guy that tries anything . . . because the next guy that tries anything may have a sawed-off shotgun under his sweatshirt. There's a 50% chance he won't, but if you like those odds, you need more help than a karate instructor can give you.

Singles Bars
(See also Computer Dating Services; Con Games; Murder; Purse Snatching; Personal Safety; Rape.)

When it comes to singles bars, there's not much left worth saying that wasn't said in Judith Rossner's novel *Looking for Mister Goodbar*. Probably the saddest part of the story was the fact that the school teacher—the victim in the story— didn't do much to help herself. She was sort of a mark. There isn't much any of us can do to make people safe who've made up their minds it's not important.

But in this society, it is important!

Fortunately, a novel isn't real life, however realistic. When I go to singles bars, I see a lot of ordinary people having fun. I also see people, particularly young and swinging women, doing a lot of really dumb things. If you happen to be one of them, you may think I sound a little corny and old-fashioned. But believe me, I've known lots of guys who could make all your worst fears come true.

Crime prevention is like an insurance policy: You have to keep the payments up. The first time you let the payments drop is when something happens to you. No matter how tired you get of reading about hardening the target and keeping your paranoia up, just remember that it's the best insurance against the possibility of crime happening to you.

In my crime prevention talks and seminars around the country, and in radio and television appearances, there's one thing I continually get back to: Never but never enter a car that you've left standing for any length of time without first checking the back seat. Do it even if the car has been locked.

My family and friends hear me say this again and again, in public appearances and in private. So what happens? Recently my stepdaughter left a singles bar in Dallas. She's heard my rap more times than anyone deserves to. But she didn't bother to look in the backseat before getting into her car. A big dude was in there waiting for her. He punched her, raped her, and nearly killed her. Now she not only avoids singles bars, she's afraid to get into her car. The point is: Don't *you* let *your* insurance policy lapse.

Most of the guys and gals who go to singles bars may be okay. The number of flakes and criminal types you're likely to run into varies from place to place. But suffice it to say, the risk is always there. It's important for you to keep that in mind, whenever you make the singles bar scene.

Here are a few dos and don'ts that may help you make it through the night. If you pay attention, you'll probably be safe, rather than sorry, the morning after.

1. Pay your own way. Don't accept drinks, drugs, or anything else from people you don't know.

2. Keep your guard up. Don't drop your guard because you're someplace you go to regularly. Bars are public places. No one can predict or control who's going to walk in the door. Even private clubs, although somewhat better controlled, can be dangerous places.

3. Age doesn't matter. Don't kid yourself that just because a person is young or in your age group he's okay. Statistically, the largest numbers of criminals are *under 30*.

4. Carry the minimum. When you know you're going to be spending your evening in singles bars, don't take along a purse or wallet stuffed with cash or credit cards. Take only what you need for the evening. Sign tabs where you can. Don't flaunt valuables like expensive jewelry, cigarette cases, or lighters.

5. Take it easy on the sauce. Order a tall seltzer and lime for the long haul. Keep your paranoia sober. It has important work to do.

6. Stay put. Don't let anyone you don't know or feel completely safe with talk you into leaving the bar and going with him or her, whether he or she is alone or with friends.

7. Stay out of cars with strangers. Don't accept offers of "rides home" from strangers. If you're driving, don't offer strangers rides.

8. Don't give out personal info. Don't share your home address or telephone number with others at the bar. If you want to see them again, make a date to meet back at the bar or give them your work or business number.

9. Hide your ID. Don't leave purses, wallets, or other ID on the bar or barstool where anyone can get at it and get vital information about you while you're dancing or otherwise occupied. Keep your possessions in your hand or in your jacket pocket, or leave them with someone you can trust.

10. Be a detective. If you meet someone you think is "special," find out about him or her *before* you agree to a car ride, a date or sex. Talk to the person's friends, if you can. Look the person up in the phone book. By hook or by crook, find out everything you can. If you draw a blank or get negative vibes from people, steer clear of the person. Don't let your emotions push you into a trap. Remember the old saying, guys (and babes) are like Corvettes, there's a new one along every minute.

Social Security Theft
(See also Elderly, Common Crimes Against the.)

Social Security check theft is a real danger for older citizens in our large cities. Most checks are stolen from mail boxes, but some are extorted by criminals who force their victims to sign the checks before cashing them. The crime becomes more serious when the elderly cash their checks and are caught in the street by muggers and purse snatchers.

USE DIRECT DEPOSIT

The surest way to short-circuit this whole process is to arrange with your bank (or if you don't have a bank now, the bank nearest your home) for direct deposit of your government checks into your personal checking account. Direct deposit, which is jointly operated by the Social Security Administration and U.S. banks, is an optional program wherein the government, upon your request, will send your Social Security payments directly to your bank for deposit in your personal checking account. As an additional benefit, many of the banks participating in the program offer free checking to people over 65.

As I point out in the section "Elderly, Common Crimes Against the," the direct deposit program removes the opportunity for crooks to get hold of your Social Security money through theft, purse snatching, mugging, robbery, or intimidation. You can pay your bills by check, shop by check, and do nearly all your necessary transactions without exposing your money to thieves.

If you're receiving Social Security benefits, live alone, and haven't already signed up for direct deposit, you should. Once the news that your Social Security checks go directly to the bank gets around your neighborhood, your life will be a lot simpler—and a lot safer. Ask your bank to give you the necessary forms to sign, or for further information write to the following: Direct Deposit, C2, Annex 1, PB-706, Washington, D.C. 20226.

WHAT TO DO IF IT HAPPENS

Both stealing Social Security checks and cashing U.S. Government checks by those other than the payee are federal offenses punishable by heavy fines and imprisonment. In addition, many states have recently enacted strict laws protecting older citizens from crime.

If you're an older citizen, you're better protected today than you've ever been. And criminals who attempt to prey on you are in more trouble than they've ever been. It remains for you to tell your law enforcement people who they are.

Here is what you should do if someone has stolen your Social Security check:

1. Call the cops. Call the police and give them the thief's name and address, if you know it. If not, give them his description and details such as where and when he usually makes his appearances.

2. Don't be afraid to report your tormentors. Don't let anyone, either criminals or your neighbors, talk you into remaining silent and putting up with the crime. Don't allow yourself to be intimidated into being a perpetual victim.

3. Ask for help. If the police aren't helpful or you're getting continued harassment from neighbors, contact the Senior Citizens Center in your community. Tell them what you're up against and ask them to help you.

4. Keep records. If possible, keep track of your Social Security check numbers. Report any that are extorted or taken from you to the authorities.

T

Telephone Security
(See also Alarms, Home and Business; Apartment Security; Burglary; House Security; Personal Safety.)

The first thing a *professional* burglar will usually do before going into your house or business is cut the phone wires. If he knows they're tied into a central station alarm (see "Alarms, Home and Business"), he may use one of several techniques to stop the alarm from sounding a warning that it's being tampered with.

The best security against an intruder cutting your phone line is to have the phone company install special, underground lines leading into your house or building through the foundation. This second line is then added to the alarm system. The burglar cuts your regular line in the belief that he's safe to move about undetected. Your special line triggers the alarm the moment he opens a window or goes through a door. If you're in the house, you still have a line to call the police.

TIPS ON PHONE SECURITY

Most of us won't attract top professional burglars; but even amateurs and kids can be a problem. Anyone who breaks into your house is likely to take the phone off the hook. It's standard procedure. From then on, your single-line phone is useless. You're at the mercy of the intruder.

Here are a few tips on how to keep your telephones secure and working for you:

1. Keep one phone hidden. The simplest, most economical form of telephone security is to restrict yourself to one phone and to keep it in your locked bedroom. While this won't protect you from the professional who cuts lines, it will stop intruders from disabling your phone by taking it off the hook.

2. Use modular systems. If it's essential for you to have extension telephones in other rooms, use one or two portable extensions with modular wall plugs (see Figure 13). Install, or have installed, outlets in the rooms where the phones are likely to be used. You can then move phones from room to room. Be sure to store extra phones in a secure place at night and to keep one set in the locked bedroom for emergencies.

3. Lock them away. At the very least, arrange extension phones so that they can be locked away in cabinets or desk drawers when not in use.

4. Try a cordless phone. An alternative to extension phones is the portable cordless, electronic phone. These units have a base that's attached to the phone line. The newest multiple handsets with touch-tone dial and other features can be used to make or receive phone calls anywhere within 700 feet. Since the phone sets work on the FM radio principle, no cords connect them with the base unit.

Cordless phones are a cinch to protect since the base unit can be securely locked away at all times and the phones themselves kept close at hand, regardless of where you are. Many cordless models include intercoms that operate indepen-

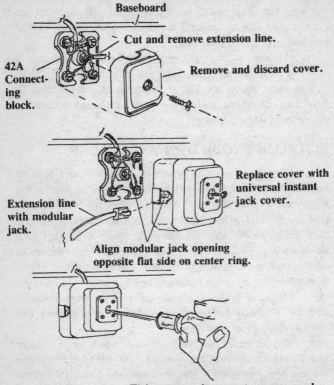

Baseboard

Cut and remove extension line.

42A Connecting block.

Remove and discard cover.

Replace cover with universal instant jack cover.

Extension line with modular jack.

Align modular jack opening opposite flat side on center ring.

Tighten screw in cover to assure good contact with 42A connecting block.

Figure 13. REPLACING EXTENSION PHONES WITH UNIVERSAL INSTANT JACKS*.

(Installation of Universal Instant Jack phone connections gives you the option of modular or 4-prong plugs on your extension phone.)

dently of phone calls, redial memories, and other special features. Cordless models with a range of 100–500 feet are currently on sale for under $75. Models with a range of 700 feet and over cost a little more.

An additional incentive to converting to cordless phones is that you eliminate equipment charges on your phone bill.

You must advise the phone company of conversation to your own equipment, and they will tell you how to return the phone company equipment.

INSTALLING YOUR OWN WALL JACKS

1. From your existing phone(s), trace the wire leading into the phone back to where it comes through the wall. If the wire leads to another phone, trace the other wire leading into that phone, and so on, until you reach the point where the wire goes into the wall. At that point, there will either be a small square metal box with a single screw in its center (a 42A block) as in Figure 13 (the most common form of phone installation) or an outlet plate and box inside the wall similar to an electrical outlet box. If your phones were installed recently, you may have modular outlets described above and can simply unplug your phones.

2. If you have the common 42A block or a similar installation where wires are attached, you will have to remove the red and green wires that lead into the block from your phone and then replace the block cover with a Universal Instant Jack as shown in Figure 13. All the necessary hardware, which includes basic installation instructions, is available from any store that sells phones, including major department stores, discount electronics stores and radio/TV outlets.

3. Whenever you replace wires, always remember that green wires are attached with green wires, red wires attached with red wires. *Never* attach red and green together on the same terminal post.

Vacation Travel Tips
(See also Airport Crime; Camping Trip Crime; Credit Cards; Hostage Situations.)

Not so long ago, you probably would have loaded up your car or hopped on a plane and gone just about anyplace you wanted to go without fear of crime or violence. But not any more. Crime statistics, both national and international, tell the story. On practically every page of this book, we talk about the risks here at home. But there are also risks abroad, and these risks are increasing across the board: more murder, more rape, more kidnapping and hostage taking, and especially more theft.

Yet, when we're getting ready for that first trip abroad in 5 years, we don't want to think about risk. We don't want to believe that the tree-lined streets, the friendly pubs and restaurants, and the good hotels have changed. But they have, just as surely as our own country has changed.

Here are some tips to help keep you as safe on your next vacation as you were on your last. Maybe safer.

THE SAFE VACATION

1. **Back to prevention.** The best way to avoid trouble is not to ask for it. Eliminate potential trouble spots from your travel plans. If the natives are rioting in one of your stopovers, change your plans. Stay away from places "in the news" because of agitation against U.S. policies or places where there are known terrorist groups.

2. **Travel light.** Pack wash and wear and permanent press clothes. Limit yourself to a large suitcase and a clothing bag for longer trips; a smaller suitcase and clothing bag for shorter ones. The less you carry, the less you're likely to lose or have stolen. Getting luggage in and out of hotels and vacation resorts is always a pain in the neck and, more to the point, time consuming. The more time things take, the more opportunity there is for theft or loss.

3. **Empty the car.** If you take your car or rent one, don't leave valuables in it. Most hotels and resorts in this country and many in most foreign countries have strictly enforced valet parking by the hotel employees. You have very little control over the security of your car or its contents. The best you can do is lock the trunk and leave only the ignition key with the attendant.

4. **Use the consulate.** If have serious problems with anyone during your travels outside the country, contact the nearest U.S. consular official immediately, then contact the local police. If you're threatened, or think you're being followed, or suspect that you or others in your party may be in danger, call the consulate. Don't shrug it off.

5. **Take taxis and guided tours.** When visiting foreign cities, use a cab or hack to get back and forth from your hotel, and do your sight-seeing with an organized tour group. Don't wander around the city alone or get into so-called gypsy cabs or private vehicles offered for hire. The days

when inexperienced travelers could explore strange cities without endangering themselves are over. Try to arrange your vacation so you don't find yourself on the streets alone.

6. Use hotel safes. Major luxury hotels and vacation resorts have safes available without charge to their guests. Use them at night to store your jewelry, expensive cameras, large amounts of cash, and traveler's checks, and to store valuables you won't need while you're out sight-seeing. Here again, prevention will serve you better than all the protection you can pay for.

Vandalism
(See also Camping Trip Crime; Car Stripping; House Security.)

With few exceptions, vandalism is done by misguided, poorly motivated, idle kids. The results are the costly destruction of public and private property. Although only a relatively small number of actual instances of vandalism are reported (in 1982 total estimated arrests were 245,700), one authoritative estimate places the cost of vandalism to Americans at $1 billion per year.

Jail sentences for the offenders are not the answer. Only a small part of vandalism can be changed by the legal system. It's up to us to prevent vandalism: first, as much as possible, by removing the opportunity, and second, by doing a better job with our own and neighborhood kids in educating them to the fact that their vandalism hurts them and their families as much as it does anyone else.

For the ones who *do* get caught, we should demand and enforce stiff probationary community work sentences to repair the damage done and impose stiff fines and civil liability in damage actions for all parents of children still living at home, regardless of age, who are convicted of vandalism.

Until such laws are passed and enforced uniformly, we'll continue to see more and more vandalism in our society. And for the laws to do any good, we must bring the parents into the picture. Let's stop giving vandals a free ride and let's get the $1 billion back from the people who are responsible. Although similar laws are already on the books and being enforced in some places, we need *uniformity* if we really want to reduce the costly waste of vandalism.

HARDENING THE TARGET

A surprising amount of vandalism happens because people are careless or lazy: We don't bring in the garbage cans after the collection. The next day we go out to get them and they've been smashed flatter than pancakes, or they're missing entirely. Or we decide that boarding up the house at the lake is too much trouble. In the spring it costs us a small fortune to do repairs and put in new windows.

Not all vandalism can be avoided, but much of it can if we simply pay more attention to the risks. In a sense, we play into the hands of vandals by not taking a few precautions. Here are a few guidelines for hardening the target.

Your Home

1. Lighting is very important. Your outdoor lighting is a key to keeping you and your property safe from predators and vandals who operate, for obvious reasons, under cover of darkness. Make sure the street and yard in front of your house are well lit. Kids who drive cars through yards and flower beds don't like to be seen doing it. In most cases, they'll pick a target with less light where the chances are less that they'll be seen or recognized.

2.   Tighten security in apartments. If you live in a high-rise apartment house plagued by vandalism, make sure your super or manager is doing his share to tighten security. He should keep all doors and gates locked at all times when not in use. If vandalism continues despite good precautionary measures, form a neighborhood watch group. Try to determine where the vandals are coming from and who the parents are, and then prosecute vigorously to put an end to the nonsense.

3.   Form patrols. In suburban and rural communities, form neighborhood night patrols, with each resident assuming his or her share of the duties. Stake out places hit the most often. Call the police. Always press charges.

4.   Protect mailboxes. In suburban and rural areas, the mailbox down at the end of the driveway is usually the prime target. Using 8 or 10 penny nails, toenail it to a two-by-six base, then toenail the base to a good sturdy tree at roadside, if there is one. If there isn't, make a metal base from angle iron purchased at your hardware store and bolt it to the mailbox, then bolt the whole works to a 2-inch pipe. Set the pipe in a few cubic feet of concrete. If you can work out the wiring, put a light in a tree or on a pole near the mailbox.

5.   Cover up cars and machinery. Don't leave cars or machinery outdoors unless you absolutely have to. If you have no place to store them, cover them and keep them as close to your house as possible.

6.   Expand your home alarm system. This will protect you from vandals as well as burglars. Now that wireless perimeter alarm systems are available, you can defend larger areas of your property without having to lay cable or string wire all over the place. With a little ingenuity you or your handyperson can include fences, gates, driveway, walks, cars, and other vehicles and machinery left outdoors, even your mailbox, in the protection of your home alarm system.

RAY JOHNSON'S TOTAL SECURITY

Other Property

1. Protect the summer place. If you have a cabin or vacation home that you leave vacant for part of the year, pay someone to keep an eye on it, preferably someone who lives nearby and has kids that know what's going on. Before you leave to return to the nine-to-five routine, double padlock doors; board up all windows; turn off the water at the street, if you have town water; turn off all electricity; and lock up pumps and other equipment.

2. Protect your car. You have some options for protecting your car. Instead of replacing your antenna every time some jerk breaks it off, install a retractable power antenna. Or have a hidden windshield antenna installed inside your car. To secure wheels, use locking lug nuts or hubcaps. As part of your car alarm, install a motion sensor that will go off if anyone fools with your car or tries to move it.

You can do a lot to protect your tires and other parts of your car from vandals by developing good parking habits. Get into the habit of parking your car in an attended lot or parking garage whenever you can. If you must leave it on the street, park it in a well-lighted place in full view of other passing cars and pedestrians. Don't hide it back behind the store or down some dark alley.

3. Use safe storage areas. Put boats, trailers, motor homes, etc., in safe storage when not in regular use. Don't leave them out and exposed to vandalism.

4. Keep yourself up-to-date on security technology. Exciting things are happening in electronics. Almost daily, new more advanced devices are coming on the market. But while sophistication is going up, prices are coming down. It's definitely worth your while to discuss your home security needs with your local electronics dealer. With a little imagination, you can make yourself a lot more unpopular with burglars and vandals. And if everyone on your block does the same, your vandals will probably move on to another part of town.

W

Windows
(See also Alarms, Home and Business; Burglary; Doors; Locks.)

Windows, by definition, are vulnerable to break-ins. As pieces of thin glass surrounded by a wood or metal frame, they're the weakest link in the home security chain. Here, especially, we need to harden the target—but how? Here are a few ideas for making your windows more secure:

NEW CONSTRUCTION: BUYING WINDOWS FOR PRIVACY AND SECURITY

As I pointed out in the section on "Doors," the best of all possible security worlds is to do it in original construction. The same goes for windows. If you're still in the planning stages of a condominium or home, discuss every aspect of home security—including window security—with your contractor. Make sure the placement and design of windows give the most privacy and security for the money.

The ideal windows for conventional houses are the double-paned, aluminum-framed type with built-in storm windows.

Manufacturers of these windows advertise regularly in the shelter magazines and the do-it-yourself home remodeling magazines, usually with listings of local dealers and distributors. More and more commonly, windows and skylights are being placed high in the walls of modern homes to provide ambient light reflected through interior glass walls and openings and greater security.

UPGRADING OLD WINDOWS

Older homes present you with more of a challenge. The old-fashioned sash windows are a cinch even for amateur burglars. They have the added vulnerability of being located low in the walls.

While you can't do anything about their location, you can greatly improve security by upgrading these old windows with new, modular windows, frames, and screens. Discuss upgrading them with your local building supply store. Be sure to take along measurements of the window openings.

Modular windows are manufactured for most conventional-size window openings. Depending on how much you want to spend, you can purchase double or single windows of various strengths and qualities with prefitted frames and screens. You simply take out the old windows and put in the new, frame and all.

ALARMS AND OTHER WINDOW SECURITY

You may not want to get into the whole business of renovating windows. Good locks and a perimeter alarm with glass protection may be as far as you want to go right now.

Your perimeter alarm system should defend your windows in two ways. First, it should warn you if anyone tampers with the glass (for example, uses a cutter in trying to get through it without breaking it). Second, it should warn you if any attempt is made to jimmy the lock and open the window itself.

Foil tape is still available for window glass alarm circuits; but it's rarely used today, because glass can be cut without disturbing it, providing the intruder an opening through which he can work to disarm your alarm system. New *window glass sensors*, so sensitive they "feel" any attempt to tamper with the glass, are now the preferred protection, even though their cost is somewhat higher.

Window locks are available in various styles. To some extent, your choice will be dictated by the design of your windows. Take a sketch of a typical window with you when you go to the hardware store to order locks. Though your windows will be of varying sizes, they should all be of the same type, unless the structure has been added to during different eras, as in the case of certain historic homes.

When ordering locks, make sure your hardware salesperson orders locks with identical cylinders. Otherwise you'll wind up with a pocket full of keys.

Don't overlook locks for the sliding glass doors. See the sections on "Doors" and "Locks" for more advice on securing sliding glass doors.

Finally, to secure windows that are particularly vulnerable and won't be used in an emergency (such as basement windows in a house) you can install iron gates or bars (see "Locks").

GLOSSARY

Bear—Challenging; difficult. Something or someone formidable.

Belt—A stiff drink; sometimes a straight shot of hard liquor, such as whiskey or vodka. Also a hard blow with a fist.

Big Ones—$100 bills.

Boogie—Move about quickly. Run or drive at high speed.

Bread—Money.

Buck—Dollar.

Bug—Annoy, antagonize, anger.

Bunco—Flim-flam; the act of obtaining money through trickery.

Bust—Arrest.

Cat—A person, usually male. Someone particularly knowledgable or experienced. Someone wiley; streetwise.

CB—Citizens band radio.

Central Station—A communications center operated by a burglar alarm company that receives signals over telephone wires from subscribers' burglar alarms.

Chop Shop—A garage or similar location where stolen cars are completely disassembled for their parts.

Clunker—An old used car.

Cold Turkey—Unassisted withdrawal from drugs.

Connection—Narcotics peddler or pusher.

Cool—Very good; fine; excellent. Also composed, calm.

Crash—Narcotics withdrawal. Sometimes used loosely to denote rest or sleep.

Creep—A despicable person; a criminal. Someone odious.

Dig—(Verb.) To understand. To like a lot; enjoy; greatly admire. Also to examine closely; stare; gaze intensely; to ogle.

Digs—(Noun). Dwelling place; living quarters; home or apartment.

Dip—A pickpocket.

Door Viewer—Peephole; bull's eye viewer; wide-angle (90 degree) viewer.

Dude—A male person, usually spectacular in dress, bearing, appearance, etc.

Far Out—Extreme; farfetched; bizarre. Sometimes used as an expression of admiration or approval.

Folsom—Folsom State Prison, Folsom, California; a maximum security prison for habitual criminals.

Glossary

Get it into Gear—React quickly; move smartly.

Gig—A job; a piece of work or employment; a robbery or stick up.

Gun Nut—Gun collector; gun enthusiast; a person with an inordinate fondness for guns and shooting.

Habit—Dope addiction.

Hit—Attack; assault; shoot; rob.

Hole—Solitary confinement.

Horn—Telephone.

Hurt—Kill.

Infrared Activated—Remote control through use of pocket transmitter.

John—Prostitute's prey; solid citizen; square john. Sometimes contemptuous for a dumb or ignorant victim.

Joint—Prison. Also a marijuana cigarette.

Kinky—Twisted personality, usually in a sexual sense. Possessing unorthodox tastes and interests.

Laid Back—Relaxed; easy going.

Lift—Steal.

Loid—A thin piece of plastic (from cellu*loid*) used to open bevel locks by insertion between bevel or spring bolt and strike plate.

Loser—A convicted felon; an unsuccessful criminal.

Make the Scene—Do it; perform; join in; attend; appear; to become part of.

Marbles—Mind (brains); sanity; rational behavior. As in to lose your marbles (go crazy).

Mark—A victim; a chosen target; an easy score (a person easily victimized).

Midnight Auto Parts—Car strippers.

Mother—A person or thing possessing some extreme quality; a monster, or monstrosity. Powerfully evil; humongous; horrific.

Nail—To strike, attack, assault, hit. Also in driving a car, to step on the gas; to accelerate.

Pad—Apartment.

Patty Cake—A harmless or timid criminal.

Perp—Perpetrator: the person who commits a crime.

Rip off—Steal from.

Sap—A kind of small club used as a weapon capable of rendering a victim senseless when struck on the head with or by it.

Sauce—Booze; hard liquor.

Scam—A scheme, usually fraudulent, for obtaining money.

Score—To accomplish successfully. The thing accomplished.

Seltzer—Seltzer water, i.e., something of little or no value.

Sharp—Smart, fashionable.

Shim—To insert a thin tool, or shim, between the door stop and door jamb in order to force bevel-type locks (also see *Loid*).

Shot Bag—A heavy canvas bag filled with lead or steel shot used as a club; similar to sap (above).

Glossary

Split—Go; leave; depart; move out.

Spring—Pay for; spend. Also release, as from jail.

Square—Legitimate; honest.

Straight—Honest, square.

Taker—Robber; burglar; thief; rapist.

Till Tapping—Stealing money from an employer; usually taking small amounts regularly from cash drawer or register.

Together—Well organized; well adjusted; a dependable person.

Turkey—A fool; a figure of fun; anyone who behaves in a bizarre fashion; a dummy; a ridiculous or tiresome person; a loser; a thief; a criminal.

Up to Speed—Up-to-date; current status.